Synthese Library

Studies in Epistemology, Logic, Methodology, and Philosophy of Science

Volume 512

The aim of *Synthese Library* is to provide a forum for the best current work in the methodology and philosophy of science and in epistemology, all broadly understood. A wide variety of different approaches have traditionally been represented in the Library, and every effort is made to maintain this variety, not for its own sake, but because we believe that there are many fruitful and illuminating approaches to the philosophy of science and related disciplines.

Special attention is paid to methodological studies which illustrate the interplay of empirical and philosophical viewpoints and to contributions to the formal (logical, set-theoretical, mathematical, information-theoretical, decision-theoretical, etc.) methodology of empirical sciences. Likewise, the applications of logical methods to epistemology as well as philosophically and methodologically relevant studies in logic are strongly encouraged. The emphasis on logic will be tempered by interest in the psychological, historical, and sociological aspects of science. In addition to monographs *Synthese Library* publishes thematically unified anthologies and edited volumes with a well-defined topical focus inside the aim and scope of the book series. The contributions in the volumes are expected to be focused and structurally organized in accordance with the central theme(s), and should be tied together by an extensive editorial introduction or set of introductions if the volume is divided into parts. An extensive bibliography and index are mandatory.

Mirco Sambrotta
Editor

Metaphysics Today: In Conversation with Amie Thomasson

Springer

Editor
Mirco Sambrotta
Department of Analytic Philosophy
Institute of Philosophy, Slovak Academy of Sciences, v.v.i.
Bratislava, Slovakia

ISSN 0166-6991 ISSN 2542-8292 (electronic)
Synthese Library
ISBN 978-3-031-98467-9 ISBN 978-3-031-98468-6 (eBook)
https://doi.org/10.1007/978-3-031-98468-6

This Springer imprint is published by the registered company Springer Nature Switzerland AG
The registered company address is: Gewerbestrasse 11, 6330 Cham, Switzerland

Acknowledgements

This volume stems from the 28th edition of the SIUCC (Inter-University Workshop on Philosophy and Cognitive Science), held in December 2019 at the University of Granada and devoted to the work of Amie Thomasson. We warmly thank the SIUCC and its organizers for providing the occasion that first brought these conversations to life. We are deeply grateful to all the contributors for their insightful and engaging chapters. Special thanks go to Amie Thomasson for her generous replies and inspiring work, which lie at the heart of this collection. We also sincerely acknowledge the support of the Institute of Philosophy, Slovak Academy of Sciences, v.v.i. in Bratislava, which was instrumental in bringing this project to fruition.

Contents

Contributors

Juan Acero Philosophy Department I, University of Granada, Granada, Spain

Otávio Bueno Department of Philosophy, University of Miami, Coral Gables, FL, USA

Daniel Dohrn Department of Philosophy, Università degli Studi di Milano, Milan, Italy

Xavier de Donato-Rodríguez University of Santiago de Compostela, Santiago, Spain

José L. Falguera University of Santiago de Compostela, Santiago, Spain

M. J. García-Encinas University of Granada, Granada, Spain

Theodore D. Locke University of Massachusetts, Boston, MA, USA

Irene Olivero DIG Department, Polytechnic University of Milan, Milan, Italy

Nemesio García-Carril Puy Department of Philosophy and Society, Complutense University of Madrid, Madrid, Spain

Pablo Rychter Department of Philosophy, Universitat de València, Valencia, Spain

Mirco Sambrotta Institute of Philosophy, Slovak Academy of Sciences, v.v.i., Bratislava, Slovakia

Amie L. Thomasson Department of Philosophy, Dartmouth College, Hanover, NH, USA

Chapter 1
Introduction

M. J. García-Encinas

The 28th edition of the SIUCC (Inter-University Workshop on Philosophy and Cognitive Science) focused on the work of Amie L. Thomasson. It was held in December 3–5, 2019 and took place at the University of Granada. The SIUCC is an annual conference consisting in three-day workshops organized under the auspices of the Spanish Society for Analytic Philosophy. Its goal is to bring the work of leading international philosophers in the most important fields of analytical philosophy to the Spanish philosophical community. Five years ago, our working group on Language and Metaphysics proposed to invite Amie Thomasson for the 2019 edition in Granada, as we thought she was a true leader in the field of metaphysics. For a year we had held a seminar based on her classic *Fiction and metaphysics*. Young researchers within the group were keen on her work. Mirco Sambrotta was then finishing his doctoral thesis on her metametaphysics. And I also remember Professor Acero used to mention her expertise in Ingarden's phenomenology. We all agreed that she was perfect for the SIUCC.

Our thoughts of excellence and high originality about Thomasson's work have been greatly surpassed by the enormous influence she has continued to accumulate since then. In 2019 she had already published *Fiction and metaphysics* (Cambridge University Press, 1999), *Ordinary objects* (Oxford University Press, 2007), and *Ontology made easy* (Oxford University Press, 2015). She had moved from Miami to Dartmouth in 2017, and when visiting Granada she was finishing her *Norms and necessity* (Oxford University Press, 2020). Her three conferences in Granada presented the main arguments in this book. Another volume about her work has recently appeared as *Thomasson on ontology* (Garcia-Godinez, 2023), and she is now about to publish *Rethinking metaphysics* (Oxford University Press). You can find more

M. J. García-Encinas (✉)
University of Granada, Granada, Spain
e-mail: encinas@ugr.es

M. Sambrotta (ed.), *Metaphysics Today: In Conversation with Amie Thomasson*, Synthese Library 512, https://doi.org/10.1007/978-3-031-98468-6_1

information about her at https://www.amiethomasson.org/. This volume rewrites the main talks and discussions at the SIUCC Workshop.

Following her phenomenological instincts, Thomasson (1999, 2003) was first in defending and introducing for the analytic tradition a new form of ontology. This includes a new category of *abstract artifacts* within which we should especially count fictional characters, such as Holmes, and social objects, such as novels, state constitutions and songs. According to Thomasson, there are artifacts that are abstract objects. They are abstract because they are intentional. And they are not Platonic, but also exist in time and contingently. Intentional objects exist, just as tables and chairs do, because they are created by people who could have decided otherwise. Abstract objects can change and be destroyed. Being fundamentally intentional objects, their nature is wholly dependent on our social and linguistic practices. They begin to exist in time, have their own identity conditions of persistence and sometimes, as in the case of novels or pictures, also in space. Same as they could have not existed, they can disappear.

The view is more than inviting. However, probably due to their volatile nature, abstract artifacts and, in particular, fictional characters when understood as abstract artifacts, cannot but rise perplexities. Thus, the volume opens with two difficulties. A very well-known difficulty is that one can truly say incompatible or even inconsistent things about many of these artifacts, as when we say that Harry Lime was a great creation by Greene and that Harry Lime was a penicillin trafficker. In his paper, **Juan José Acero** presses Thomasson on this hard point arguing that, on her view, 'Harry Lime' should be the name of an artifact; so it cannot be the same name on both occasions. She states back that the same name 'Harry Lime' can have different sets of application conditions and still be the same name because its uses are clearly related. Both Acero's paper and Thomasson's reply show that solving this issue is not an easy task. Another difficulty for the realism of abstract artifacts is raised by **Otávio Bueno,** within the context of films. Among other things, he asks whether fictional characters can be seen when watching a film. Considering that Thomasson argues that they are real, her view could contribute to an answer in the positive. However, given fictional characters are also abstract, how could they be seen? Difficulties are various, and Bueno opts for an empiricist alternative where the existence of fictional characters is not at issue. Thomasson's suggestive reply is to claim for a quasi-perception of fictional characters, also in films.

Her pragmatist view that names correspond to sets of application conditions and that meaning is dependent on linguistic practices signified a new form of doing metaphysics and a novel approach within the field of metametaphysics. For starters, Thomasson (2007a, 2015) denied that metaphysics is a question of discovery. Reminding the qua-problem, she argues that if there are many distinct things that one could aim at referring to by a given term, disambiguating among them will always require background assumptions about what kind of thing the term refers to. Thus, existence questions have sense only within a conceptual frame. Moreover, Thomasson (2016) argues that external metalinguistic negotiation, and choosing among conceptual schemes, is what we are finally called upon to do in order to achieve a better understanding of our world and its changes. **Nemesio**

García-Carril, in his contribution within the general frame of the metaphysics of art, invites us to consider whether this general view can in fact satisfy our metaphysical and aesthetical needs, or we demand a more "substantive" ontology for art.

Mirco Sambrotta's paper confronts the question of our ontological and existence expressions. In an attempt that goes further than Thomasson's general view in *Ontology made easy* (2015), he claims that ontological judgements actually express our commitments to the adoption of a conceptual structure and certain forms of language. In response, Thomasson insists on her Carnap-inspired distinction between internal and external questions about existence. Internal questions are easily responded in the affirmative, and external questions are not about factual existence but show disagreements about the linguistic or conceptual scheme that we should adopt.

This distinction between internal and external ontological questions is key in **Xavier Donato and J. L. Falguera**'s paper. They argue that Thomasson's views can be extended to debates on scientific ontology in the following sense. They start from Thomasson's thesis that existence questions are always internal to the theories so existence questions are easy to answer in the affirmative. They apply this thesis to scientific theories and argue, first, that when a community accepts a scientific theory, the use of theoretical and observational terms, and the quantification over variables related to theoretical and observational entities, is internal to the theory. Therefore, theoretical scientific entities can be said to exist, but just internally. And, second, that in this sense theoretical scientific entities can be understood as existing abstract artifacts.

Pablo Rychter also discusses Thomasson's view on easy ontology, but he centers his discussion on her defense of the easy existence of facts (Thomasson, 2007b). According to Thomasson (2007b), if nine is an even number, then there are numbers. In the same way, if this was a white page in the past, it then follows that there is the fact that this was a white page. Facts are easy entities resulting from trivial inferences from sentences declaring the instantiation of a property by an object, thus existing in an easy way, internally. Rychter disagrees with this picture. However, he also argues that facts do not exist either in the more robust and explanatory sense of Armstrong. Rather, he claims, facts are merely pretended to exist.

That norms and rules and language license us to introduce reference to new objects and kinds of objects, and to acquire knowledge about them, is one of Thomasson's main theoretical proposals. As shown by the two papers I just introduced, this is a thesis of wide scope, so it can be widely applied. In different places, Thomasson (2007b, 2013, 2020) intelligentlly applies this thesis to modality. Modal claims, such as that iron cannot be transparent, are about semantic rules governing the semantics of 'iron' and 'transparency' or are made true by semantic conventions. **Daniel Dohrn** interprets Thomasson's modal normativism as the theory that modal statements about metaphysical possibilities and necessities are object-level statements of semantic rules. This theory, on his view, precludes substantial metaphysical enquiry. But see Thomasson's complaints in reply.

More in accord with Thomasson's modal theory, **Theodore Locke** considers and rejects the possible thought that modal normativism cannot come with literal talk about possible worlds. Not only is it compatible with modal normativism that there are possible worlds, but possible worlds can also work as illustrations of the semantic rules, and the consequences of these rules, that modal talk conveys.

Also within the general framework of Thomasson's metaphysics and modal normativism, thus also taking clear distance from classic metaphysicians, **Irene Olivero** argues that disputes about artifactual kinds, just like those about natural kinds, are to be understood as metaphysical negotiations about how the relevant terms ought to be applied. They are not disputes about how terms are actually applied, nor inquires about real natures that determine the extension of these terms.

In general, all papers in the volume pay tribute to what I take to be Thomasson's most original, controversial and thought-provoking contribution to metaphysics today, that is, her proposal that the *nature* of our ontologies is determined by our beliefs and our social and theoretical practices.

I said above that this volume rewrites the main talks presented at the SIUCC 2019 on Thomasson's work. For reasons beyond the control of all of us who contribute to the volume, its edition has been delayed. But we are as proud as we were at the time of the conference to have been able to participate in this project. It was then a pleasure to meet Amie. In addition to demonstrating why she deserves the position she now occupies as one of our best contemporary thinkers, I remember her as a very honest, intelligent and attentive person. The fact that she has carefully and kindly commented on each paper in this volume is proof of this and highly speaks of her. It has also been my pleasure now to recall those three intense days.[1]

References

Garcia-Godinez, M. (2023). *Thomasson on ontology*. Palgrave.

Thomasson, A. L. (1999). *Fiction and metaphysics*. Cambridge University Press.

Thomasson, A. L. (2003). Fictional characters and literary practices. *British Journal of Aesthetics, 43*(2), 138–157.

Thomasson, A. L. (2007a). *Ordinary objects*. Oxford University Press.

Thomasson, A. L. (2007b). Modal normativism and the method of metaphysics. *Philosophical Topics, 35*, 135–160.

Thomasson, A. L. (2013). Norms and necessity. *Southern Journal of Philosophy, 51*, 143–160.

Thomasson, A. L. (2015). *Ontology made easy*. Oxford University Press.

Thomasson, A. L. (2016). Metaphysical disputes and metalinguistic negotiation. *Analytic Philosophy, 57*(4), 1–28.

Thomasson, A. L. (2020). *Norms and necessity*. Oxford University Press.

[1] The 28th edition of the SIUCC was co-organized by Research Project FFI2015–65953-P, funded by the MINECO, and by the Spanish Society for Analytic Philosophy.

Chapter 2
Fiction, Artifact and Intentionality

Juan Acero

Abstract Let us focus on the following argument:

1. Harry Lime deals with adulterated peniciline in the black market
2. Harry Lime is the unexpected character in The Third Man
3. The unexpected character in the Third Man deals with adulterated peniciline in the black market

Is this inference sound? A theory on the nature of fictional characters laid by Amie Thomasson has the consequence that the truth of (**1**) and (**2**) does not guarantee the truth of (**3**). The reason would be that what looks one proper name, 'Harry Lime', is either two different names or one name that is used in one way in (**1**) and in another way in (**2**) and (**3**). No matter what route is chosen, (**1**) and (**2**) do not provide the conditions which made it possible for the sustitutivy of identity principle to deliver (**3**). This diagnosis is based on the fact that the occurrence of 'Harry Lime' in (**2**) and (**3**) is used to refer to an abstract entity while the occurrence of 'Harry Lime' in (**1**) would only refer to a character in the context of The Third Man's plot. Whoever thinks it necessary to distinguish between what is narrated in a plot and what critics and sophisticated commentators remark on it seems to be forced to reject the inference from {(**1**), (**2**)} to (**3**). The words 'Harry Lime' do not make the same semantic contribution in each premise of the inference. I will to resist this line of criticism by giving a chance to an alternative approach that has that there is only one proper name in (**1**)–(**3**) and it keeps its meaning constant through its constituents. The main idea is that The Third Man's script authors—Graham Greene, Alex Korda and Orson Welles—simply imagined a character and had to introduce the name to put up **a plot.**

J. Acero (✉)
Philosophy Department I, University of Granada, Granada, Spain
e-mail: acero@ugr.es

M. Sambrotta (ed.), *Metaphysics Today: In Conversation with Amie Thomasson*, Synthese Library 512, https://doi.org/10.1007/978-3-031-98468-6_2

2.1 The Inference

A and *B* are chatting about novel and film characters. *A* mentions some Harry Lime and B asks A who he is. *A* answers by uttering the statement[1] (1), after which *B* shows an interest on what Harry Lime is like. *A* now utters (2) and *B* infers (3) thereafter:

1. Harry Lime is the character created by Graham Greene in *The Third Man* whose role was played by Orson Welles in Carol Reed's film.
2. Harry Lime traffics in diluted penicillin.
3. The character created by Graham Greene in *The Third Man* whose role was played by Orson Welles in Carol Reed's film traffics in diluted penicillin.[2]

Is this inference logically sound? In accepting that (1) and (2) are both true, should the laws of logic force us to accept that (3) is also true? (In what follows I will call the inference from (1) and (2) to (3) the Inference.)

According to Amie Thomasson's viewpoints on fictional characters, the Inference should be unsound, because the proper name 'Harry Lime' does not refer to[3] the same item in (1) and (2). The contrary judgment, i.e. that the same individual is pointed at in both premises, is not without any backing—and it is the one I will favor. Five targets will be aimed at from now on. First, I will set out Thomasson's views on fictional characters and suggest an alternative to her explanation of why the Inference gets stuck before arriving at its conclusion, an alternative that hinges on imagination and the intentional nature of its acts. Then I will disable some arguments of Thomasson's that hinder the consolidation of that alternative. Third, I will put on display two principles that help shed light of the affinities and differences between Thomasson's stance and mine and, finally, argue that she does not detect a deeper difficulty—the Equivocality Problem—that traces out a dividing line between both approaches. The relationship of these principles to the Equivocality Problem will stand out at the end of the journey.

The discussion of Thomasson's views will be mainly restricted to those she expounds in *Fiction and Metaphysics* (Thomasson, 1999). Since the publication of that notable work she has abandoned a substantial part of her former conceptions and followed a path that owes a lot to philosophers of quite a different profile.[4]

[1] By 'statement' I will any declarative sentence that can be used to make, in tandem with the context, an assertion. I will put aside the proper meaning of the term, namely the speech act of stating such-and-such.

[2] Harry Lime is the main character Carl Reed's film *The Third Man*, which faithfully follows a Greene's script ordered by Alexander Korda. The novel was published in 1950, whereas the film had been released the previous year.

[3] I will use 'to refer (to)', 'to name' and 'to designate' as if they were equivalent. If a name *N* refers to an individual *x*, I will say say either that *x* is *N*'s referent, that *N* names *x* or that *x* is *N*'s *designatum*.

[4] See Johnston, 1988; Schiffer, 1994, 1996, 2003; Searle, 1979. Thomasson's later views are developed in Thomasson, 2007, 2008, 2009, 2013, 2015. Her book *Ontology Made Easy* (Thomasson, 2018) provides a systematic exposition of her ontological deflactionism.

However, in order to keep the length of this chapter under control, I will refrain myself to examine her more recent developments. Her drift from stands in which some metaphysical—specifically, phenomenological—doctrines took on a central role to the deflationist stand that prevails in the Easy Ontology Approach reveals something of the atmosphere of nowadays philosophy.

Before starting out, I must introduce a conceptual note to guide readers' gazes. In dealing with fiction characters I will never have in mind character *types* like the Frustrated Wife, the Villain, the Doubtful Guy, and so on. I will take into account fictional *individuals*, viz. people, animals, fantastic beings and so on. Not to the Villain, but to Harry Lime; not to the Secret Agent, but to James Bond; not to the Resourceful Hero, but to Ulysses. I know that the difference between types and individuals so conceived is gradual: that James Bond could be thought of as a type. Nevertheless, the distance that separates types from individuals—or between schematic and enriched types, if you prefer—is beyond discussion and I will have the benefit of it.[5]

2.2 The Artifact Theory

Among the Thomasson's proposals I will address in what follows are these:

[T1] Fiction proper names have got two kinds of use: their use in the context of fiction and their use in the real context. (Put it differently, its textual use and its metatextual use.[6])

[T1] is adduced to have the analysis go on and claim that while in (1) the fiction name 'Harry Lime' is used in the real context, in (2) in the context of fiction.

[T2] In using a fiction name (e. g., 'Harry Lime') in the context of fiction, the name does not refer to anybody, that is, to no one real (actually existing).

[5] My route is not the one followed by Peter Lamarque, since I identify fictional individuals with enriched types. He would reject such a step. See Lamarque, 2003b.

[6] In the bibliography on the philosophy of fiction a different terminology is employed. The alternative does not focus o kinds of context, but on kinds of *use*. On the one hand, there is the *textual* use—the one those names have in the very fiction work's text, thanks to which those characters' life and deeds re narrated. The textual use must be distinguished from the *metatextual* use, namely the one those names are given in the statements that critics and readers resort to in dealing with the fiction work. This distinction would allow us to contend that 'Harry Lime' has a textual use in (2) and a metatextual use in (1). That both terminologies fit to each other seems to follow from the remark that in the discourse that belongs to the real context those names have got a metatextual use, whereas in the discourse that belongs to the context of fiction those names have got a textual use. We might replace the term 'discourse in the context of fiction' by 'textual discourse'; and replace the expression 'discourse in the real context' by 'metatextual context', thus bringing those two terminologies nearer. Cf. García-Carpintero, 2016, p. 68; 2019, p. 74. The alternative terminology, as well as the one employed in notes 8 and 19, comes from Bonomi, 2008.

Thus, when in Greene's novel Major Calloway says to Holly Martins: 'Harry Lime traffics in diluted penicillin', the fiction name names nobody. Moreover, there is no referential speech act through which Major Calloway refers to some Harry Lime. There being no Harry Lime, there is no naming act *within* the context of fiction. For the same reason, our speaker *A* utters (2) on witnessing the plot's unfolding. Unlike fictional characters in the real context, characters *within* fiction are not items involved in the reference relation.

[T3] Fiction proper names used in the real context refer to artifacts, e. d., they name abstract, though actually existing, entities.

Platonic entities are timeless, whereas fictional characters have got a date of issue: they are intimately linked to daily people, objects, places, situations, events or states of affairs because of their dependence on texts, images, film recordings and further vehicles imagination acts made up by their authors and promoters.

[T4] Fiction characters are entities —artifacts—created by their authors by means of particular acts of their imagination and endowed with identity and persistence conditions of their own.[7]

Fiction characters have got a date of issue. They exist from the very moment in which they are created and stay in existence as long as there remains one copy of the text or recording in which their names appear or any train of thoughts and memories in which their identity is kept up. Hence fiction characters are *dependent* entities. The fact that they exist is due to creation acts which shape their identity and continuity conditions. Their existence depends on texts, recordings or thoughts' existence, perspicuity and integrity. In order for a fiction character to remain in existence, acts are needed that so warrant it.

A literary work as such can exist only as long as there are some individuals who have the language capacities and background assumptions they need to read and understand it. If all conscious agents are destroyed, then nothing is left of fictional works or the characters represented in them but some ink on paper. Similarly, if all speakers of language die out, with the language never to be rediscovered, then the literary works peculiar to that tongue die out as well (Thomasson, 1999, p. 11).

[T5] In using a fiction name in the real context, critics, literary agents, readers, spectators, etc. refer to the fictional characters, viz. the abstract artifacts created by the authors.

Readers and spectators do not refer to the individual which supposedly features in the story plot when they put its name to use. Since they do not do it, provided that there exist nothing the name designates, how is it possible to talk about a fictional character?

[7] Although Thomasson's [T4] follows Kripke, 2013 and van Inwagen, 1977, [T5] means her entry into a new territory. In addition to this, *Fiction and Metaphysics* is deeply indebted to Roman Ingarden's ideas on literary works. See Ingarden, 1968, chap. 1.

[T6] Agents of the real context refer to an artifact when they use a fiction name, because they are actively involved in causal and dependency chains of reference transmission, chains which go back either to a sort of initial baptism or to an setting up an informal name introduction.

In other words: when a critic tokens (1), the critic gets connected to a reference transmission chain that goes back to a situation in which an author confers a name to character. The baptismal ceremony conjures up a story that imposes that name on him or her in accordance with a more or less conventional literary creation technique.

In what follows I will centre upon three of these six claims: [T2], [T3] and [T6]. I am in two minds about [T1]'s truth, but I will not now air my doubts about it.[8] I do accept [T4], being Thomasson's choice the term 'artifact' my only sting, and [T5], although I would suggest an argument trajectory to arrive at this claim somewhat different from hers.

Before finishing my presentation of the Artifact Theory, I think it necessary to lay out two assumptions that substantially contribute to its scaffolding. The first of them sustain the claims [T1], [T2 and [T3], specially the claim that 'Harry Lime' seem to be used in two different ways; even more, that under the same phonetic or graphic cloth 'H-a-r-r-y-L-i-m-e' two names, either doing its duty. Such a conclusion is based on the following

Principle of Equivalence between Reference and Existence [= *ERE*]
(For each proper name *N*) '*N*' refers if, and only if, *N* exists.

On the one hand, the use of 'Harry Lime' in (2) belongs to the context of fiction; therefore, as *ERE* demands, it names nobody. On the other hand, the use of what looks at first sight to be the same name, 'Harry Lime', in the real context does refer to something and, as a consequence, there is something, i.e., Harry Lime, that do exist. *ERE* is thus obeyed. It is natural then not only to think that the two occurrences of the expression are tokens of different types—that is, that two different names are involved but to feel the need for making it clear the nature of the *designatum* wherever there is one. When this train of thought is followed, the two occurrences of 'Harry Lime' do not refer to the same thing, and unless proper names are indexical expressions—an option ignored by all participants in the discussion—two consequences follow: that the Inference involves two different names are involved and that it is unsound.

For those aiming at defending the Inference's soundness their first reaction is to reject *ERE*, It is true that this choice has been frequently adopted in the semantics

[8] Since (2003a, 2003b) Thomasson adds to the context of fiction and the real context a third kind of context: the *internal* context. The context of fiction does not loose its name, but the real context is now called the *external* context. These changes result from noticing that a statement like (2) might be understood not only as (part of) a discourse that belongs to the context of fiction—Major Calloway would insist on lime's being a criminal, but also as (part of) a discourse that would makes plain what the text or the film set forth. The status of (1) does not change in Thomasson's new catalogue. This three-fold classification runs parallel to García-Carpintero's textual/paratextual/metatextual classification. Cf. García-Carpintero (2016: 68; 2019: 74).

of fiction research.[9] Nevertheless, it will not be the one I take. Therefore, I must explain how is it possible to reconcile three *desiderata*: (i) to abide by *ERE*; (ii) to accept that in order to bestow form and content on a literary work or a film, authors imagine characters that can in turn be objects of critics and readers' thoughts and words; and (iii) to argue that those characters do exist, although not in the way abstract entities exist. It is true that *ERE* is saved from my criticism of Thomasson's view, but I have to pay the prize of freeing the analysis of fiction from the constraints set by widely shared views in the theory of meaning, specifically widely accepted views set in the theory of reference.[10] The route that overcomes these three challenges will be judged as revisionary, for it pictures a relationship between ontology and semantics, viz. the theory of reference, that is in conflict with the one stated by a second principle:

Principle of Ontology's Accountability on Semantics [= *OAS*]
The ontology adopted in fiction discourse should be accountable to semantics' categories and rules.

OAS follows from what Jackendoff has called the Grammatical Constraint:

The *Grammatical Constraint* says that one should prefer a semantic theory that explains otherwise arbitrary generalizations about the syntax and the lexicon. [...] The point of the Grammatical Constraint is only to attempt to minimize the differences of syntactic and semantic structure, not to expect to eliminate them altogether (Jackendoff, 1993, pp. 14 f.)

Thomasson honors both *OAS* and *ERE*. *OAS* requests us to resort to the right theory of reference, and she chooses a theory that supports her conviction that fiction names appearing in the real context comply with *ERE*—they refer to artifacts. On her view, ontology, far from floating free, must follow the theory of reference she adopts, thereby subordinating her ontological preferences to semantics:

Most importantly, postulating fictional characters as abstract entities and analyzing discourse about fictional and real objects in the same way enables to offer a better theory of language than those theories that attempt to avoid fictional objects at all costs (Thomasson, 1999, pp. 113 f.).

---

[9] Classic works are Parsons, 1980; Zalta, 1983; Crittenden, 1991; Salmon, 1998.

[10] By a theory of meaning of a natural language I understand any theory that recursively assigns semantics values to the language's names and predicates, and truth-conditions to its sentences as a function of (i) those names and predicates' semantic values and (ii) their syntactic composition. Tarski-style model theory is the paradigm of such a way of articulating the theory of meaning, and one of its requirements becomes specially important, namely the idea of a universe of discourse over which names and predicates pick up their respective semantic values. On the assumption that 'Harry Lime' is an English proper name, a theory of meaning for English has to handle in parallel to how the theory deals with other proper names ('Everest', 'Empire State Building', etc.), viz. by assigning it a semantic value—an entity. Semantics as currently developed takes a fictional name to be a referential expression and has to provide a *designatum* to it. Once this last step is taken, semantic analysis crosses the ontological boundary and must decide to what kind of entities the posited *designata* belong to.

However, I will argue that vindicating the Inference's soundness is compatible with denying the claim that fiction names refer either to artifacts. Whether this maneuver also brings about the violation of *OAS* is something that has to be discussed.

2.3 The Equivocality Problem

I share a not insignificant part of Thomasson's proposals, because it seems to me that they shed light on the metaphysics of fictional entities to a higher degree than many of its alternatives'. If I am going to pursuit a different approach, that is due to the fact that Thomasson's views leave unsolved a problem that usually goes unperceived, namely whether the treatment which fiction names are subject to should be the same irrespective of their context—fictional *vs.* real—of use. Why do we make use of what looks like being the same name, e. g. 'Harry Lime', both in the context of fiction and in the real context? How is it possible that one and the same name be used, provided that in the context of fiction there exists no entity to be referred to by it—and *ERE* then prescribes that the name names nothing—whereas in the second case the name has got a referentr? According to *ERE*, the uses of the name in both kinds of context are so different that it is as if the name had been split into two, each having a semantic function of its own. The Inference's first premise requires to point at Harry Lime, e. g. Holly Martins' old friend the second, to something created by Graham Greene and Carol Reed. He would have gone from pointing at a non-existent friend of a non-existent Holly Martins to referring to an artifact—and artifacts have no friends; they are abstract. The name 'Harry Lime' seems to be equivocal, thereby forcing Thomasson to face the *Equivocality Problem*. My point, nevertheless, is that Thomasson overlooks it.

It has to be taken into account that the Equivocality Problem is not resolved, as Thomasson appears to believe, by adducing that (2.1) represents the meaning of (2):

(2.1) In *The Third Man*, Harry Lime traffics in diluted penicillin

The fact that the name 'Harry Lime' is in the scope of the operator 'In *The Third Man*' determines that its semantic, that is referential, contribution is not the same as the one the name makes in the statement (1). This remark, to which repeatedly Thomasson adheres, would led to conclude that the Inference is unsound. It can be replied that it does not have the intended consequences. Such an analysis of (2) takes 'In *The Third Man*' to be either a modal or an epistemic operator, which opens the way to reveal that (2)'s truth-conditions are those made explicit in (2.1.G):

(2.1.G) (2.1) is true in @ if, and only if, for any possible world *w*: if the story narrated in *The Third Man* is true in *w*, then the Harry Lime in *w* traffics in diluted penicillin in *w*.[11]

[11] This is the starting point of Lewis' theory of fiction. See Lewis, 1983.

(@ is the real world.) Since there is no guarantee that the Harry Lime in *w* is the same as the one referred to in uttering (1), the truth of the Inference's premises does not imply the truth of the Inference's conclusion. Therefore, (2.1.G) discloses that (2.1) expresses a *general* proposition, for it does not handle 'Harry Lime' as an expression which directly designates an individual, but as one that expresses a general concept.

The explanatory virtue of this sort of analysis is rather restricted, however. Among all its disadvantages I will briefly mention two, before arguing that no escape route from the Equivocality Problem is open for Thomasson. The first weakness which undermines (2.1.G) becomes apparent as soon as we focus on its most elaborated development[12]:

(2.1.G.N) (2.1) is true in @ if, and only if, for any possible world *w* in which the common overt beliefs of @ are true there is a possible world *v* in which a narrator tells the *The Third Man*'s story as known fact and in which Harry Lime traffics in diluted penicillin, and *v* differs on balance, less from *w* than does any other possible world *u* such that a narrator tells the *The Third Man*'s story as known fact in *u* but Harry Lime does not traffic in diluted penicillin in *u*.

Now the idea is that among all the possible worlds in which the overt beliefs shared by the author and the readers/spectators, (2.1) is true in the closest one in which the narrator tells the story as a known fact. The author makes-believe as if a narrator is acquainted with the characters' deeds and convey them while sharing, as much as it can be, the common overt beliefs shared by the author on his or her potential addressees. When the analysis takes this direction an obvious drawback comes to mind: (2.1.G.N) naturally squares with those fiction works in which a narrator is one among the rest of characters; or with works in which the author confers the narrator a testimonial role. There are many works that meet this condition. *Don Quixote* and *Moby Dick* are cases in point among literary texts; *Out of the Past* (Jacques Tourner, 1947) and *The Goodfellas* (Martin Scorsese, 1990), among films. However, there also lots of works that do not match the rule. When a figure is postulated as letting the readers or spectators know a chain of facts without earning his living in that chain, he should not be recognized as a narrator. Why should the author choose someone to tell a story instead of doing it him or herself?[13]

There is a more technical objection to be faced in postulating the implicit presence of a modal operator: the expression 'In *The Third Man*' is much better analyzed as a *circumstance-shifting* operator than as a modal one (Recanati, 2000, Parts II, III, V):

[12] The reasons that lead to this proposal constitute the heart of Lewis' essay. See Lewis, 1983, pp. 268–73.

[13] A second problem for what could be called the Modal Operator Objection is brought up by the claim that 'Harry Lime' is in 'In *The Third Man*''s scope. To hold that *proper names* rarely are in the scope of modal operators is a counter-claim that, in addition to having a lot in its favor, undermines Thomasson's criticism. See, for example, Recanati (2000, Part III).

(2.1.S) (2.1) is true if, and only if, relatively to a real-world situation (e. g. one in which the speaker notices how extreme are the difficulties that Greene's characters confront), the circumstances narrated in *The Third Man* sustain, viz. make true, that Harry Lime traffics in diluted penicillin.[14]

(2.1.S.) means that (2.1) is true if, and only if, Harry Lime—not the abstract entity, but the character that moves back and forth in post-war Vienna, brought up in Greene and Reed's imagination—traffics in diluted penicillin in those equally imagined circumstances. Alternatively, if, and only if, the circumstances narrated in *The Third Man* support that Harry Lime traffics in diluted penicillin. The operator 'In *The Third Man*' (generalizing, the operator 'In fiction *f*') brings about a world change: from @, the world of which both the speaker and the novel or the film belong to, to the circumstances that these works depict. The statement (2.1) is true, and (2.1.S) shows why this is so. It is interesting to put on record that, if (2.1.S) is on the right track, then (2) expresses a *singular* proposition (cf. Recanati, 2000, § 15.5).[15] Following Walton (1999), Recanati favors (2.1.S). He thinks that fictional names refer to individuals in the circumstances depicted in *The Third Man*:

> [...] we pretend that the world contains not only real objects and situations, but also, located in some extra dimension, imaginary objects and situations. On this view, although [Harry Lime] is an imaginary person who does not really exist, still *there is* such a person, in the imaginary realm. Since there is such a person, we can refer to him. [...] The speaker who says [in *The Third Man* that Harry Lime traffics in diluted penicillin pretends that the universe contains not only [...] the fiction authored by Graham Greene], but also ('in' the fiction in question) an imaginary individual, [Harry Lime], who does not exist 'outside' that fiction (Recanati, 2000, p. 220).[16]

I do endorse part of what these words convey. Although I prefer to avoid the verb 'to pretend' and use 'to imagine' instead of it[17]—additionally, the metaphor of an

---

[14] This is an analysis in Austinian semantics. See Recanati (2000, chaps. 5, 7, 15). The general form that the theory assigns (2.1) is the following:

(*) $[s'] \models_{@} \ll s \models_{w} \sigma \gg$

An alternative general form to (*) is (**):

(**) $[s'] \models_{@} \ll R \models_{@} \ll s \models_{w} \sigma \gg \gg$

(**) means that the situation that concerns the speaker (s') sustains a situation in which the work (novel, film, etc.) as a real-world entity (*R*) sustains the situation that in the circumstances narrated in the work (*w*) sustains that σ. According to whether we follow (*) or (**), (2.1) is respectively dealt with as in (2.1.S.i) or (2.1.S.ii):

(2.1.S.i) [The situation which concern the speaker] $\models_{@} \ll$ [the circumstances narrated in *The Third Man*] $\models_{w}$ Harry Lime traffics in diluted penicillin$\gg$

(2.1.S.ii) [The situation which concerns the speaker] $\models_{@} \ll$ *The Third Man* $\models_{@} \ll$ [the circumstances narrated in *The Third Man*] $\models_{w}$ Harry Lime traffics in diluted penicillin$\gg\gg$

[15] Two constraints, the Semantic Innocence and the Iconicity Constraints, lie behind (2.1.S). See Recanati, 2000, Part I).

[16] This insight can be elaborated in different ways. See, for instance, Deutsch, 2000; Recanati, 2000, chap. 15; Yasigawa, 2001; Howell, 2015.

[17] I do not share Walton's insight that to imagine it to pretend or to make-believe (Walton, 1978, 1999); Currie (1990); García-Carpintero (2016), as well as many others. I only welcome the inverse relation, namely, that to pretend that so-and-so is roughly to imagine that so-and-so. I sim-

extra dimension makes me uneasy, the truly important thing to notice, however, is that Recanati says that imaginary individuals only exist in fiction: that they only exist in imagination acts. Therefore, his alternative to the Modal Operator Analysis of 'In *The Third Man*' treats fiction names in a way whose conformity with *ERE* is overridden. What meaning can be found in the thought that Harry Lime does not exist *and* that there is such a person?

Unlike Recanati's analysis, Thomasson's Theory of Artifact does not have to confront this question for it keeps to *ERE*. Its weakness becomes apparent when the Equivocality Problem comes on stage. To get the right idea of what he dialectics involved is, three alternatives must be pondered: those provided by the Modal Operator (M) Analysis, the Circumstance-Shift Operator (CS) Operator, and the Artifact Theory (AT). (I) Let us suppose that we abide by (M). It follows that the fictional name 'Harry Lime', as used in a context of fiction, refers to a possible, non-actualized individual. A standard utterance of (2) illustrates the case at hand. As long as the statement (2) is understood as incorporating the operator 'In *The Third Man*', (2) has to be semantically evaluated relatively to a possible world *w*. The evaluation involves the name 'Harry Lime"s being assigned an individual of *w*'s universe of discourse. More precisely, a possible *concrete* individual. However, on looking to (1), it seems as if we should derive a different conclusion, on pain of declaring the Inference unsound. The reason is that no modal operator is need to convey (1)'s meaning, e. g. to identify the proposition that (1) expresses. In spite of it, if (1)'s 'Harry Lime' names the same as (2)'s 'Harry Lime', the conclusion to be expected is that in @ ($@ \not\approx w$) 'Harry Lime' refers to a possible, indeed a real, concrete individual. Thus, if 'Harry Lime' designates somebody in @, the individual referred to should be a *concrete* one. (II) So onto (M). In uttering (2) the speaker uses 'Harry Lime' to refer to an imaginary individual. Since the operator 'In *The Third Man*', implicit in the statement (2), is a circumstance-shifting operator, to semantically evaluate (2) we must assign the proposition expressed by the statement a truth value relatively to the circumstances that *The Third Man* depicts. However, if the name 'Harry Lime' contributes something to such a proposition, its contribution should be an individual belonging to the real world @. In fact, it contributes an imaginary, not a real, concrete individual. Since this individual has got the properties attributed to him in Greene's novel and Reed's film, (2.1) is true. On the other hand, deprived from the operator (2) is false. What matters, however, is that both in (1) and (2) 'Harry Lime' designates the same individual. (III) Let us, finally, turn to AT. 'Harry Lime' is used in (2) to refer to a possible *concrete* individual. On the other hand, AT prescribes that in the real context refers to an individual. One would

ply cannot take in the idea that Greene and Reed have made-believe that Harry Lime traffics in diluted penicillin, that he hides in the Russian sector of Vienna, and so on. It seems to me that what feeds such a view is the tempting analogy that real people are related to fiction characters like acts of committing to saying the truth are related to acts of pretending. This is not an analogy I would give support to, because I do not think that either the truth or falsity of statements is built into the constitution of fiction. As for this, I agree to the arguments in Deutsch, 2000; Lamarque & Olsen, 1994, chap. 4; Lamarque, 2003a; Moyal-Sharrock, 2009.

Table 2.1 Possible *designata* of 'Harry Lime'

	Modal operator	Circumstance-Shifting operator	Artifact Theory
Same individual	Yes	Yes	No
***Designatum* in (1)**	Possible concrete individual	Imaginary individual	No possible concrete individual
***Designatum* in (2)**	No possible concrete individual	Imaginary individual	Abstract individual

expect, in parallel to what has been spelled out in (I) and (II), that this individual has to be a *concrete* one. The reason was nicely advanced by Lewis in his classic paper:

> For one thing, is there not some perfectly good sense in which Holmes, like Nixon, is a real-life person of flesh and blood? There are stories about the exploits of super-heroes from other planets, hobbits, fires and storms, vaporous intelligences and other non-persons. But what a mistake it would be to class the Holmes stories with these! Unlike Clark Kent et al., Sherlock Holmes is just a person—a person of flesh and blood, a being in the same category as Nixon (Lewis, 1983, pp. 261 f.).

Although I would suggest that Harry Lime also belongs to the same category as Nixon, this is not what AT gives in. It provides, instead, an *abstract* individual as the name's *designatum*, not a concrete one. This combination amounts to saying that 'Everest'—the very same expression—is the name both of a mountain and, e. g. of an even natural number. I am afraid that this is utterly wrong. Let us imagine that '4' were used to name the mountain Everest; we would not still claim that the number 4 and the mountain share the same name. Instead, we would naturally say that two different names are involved. The Modal Operator approach meets the requirement that although fiction proper names' *designata* can differ in moving from a possible world to another one, their ontological categories have to remain fixed. As for the Circumstance-Shift Analysis, the compliance is obvious for the *designata* are the same. The requirement warrants the conclusion that, as far as (I) and (II) are concerned, the name 'Harry Lime' is the same in (1) and (2). The Artifact Theory cannot, on the contrary, lead to this result. It is forced to conclude that the occurrences of 'H-a-r-r-y-L-i-m-e' in (1) and (2) are occurrences of different names. (See Table 2.1.)

There is one more possibility to bear in mind: that in its canonical uses (2.1) does straightforwardly conveys the very same content as (2). Thomasson cannot afford to follow this argumentation line, since doing it would force her to recognize the Inference's soundness. The Equivocality Problem is that of explaining why 'Harry Lime' are not *two* different names in disguise. This is a problem for Thomasson as well as for all those who argue that in the real context a fiction name refers to an abstract entity.[18]

[18]The analysis favored in van Inwagen, 1977, Kripke, 2013 and García-Carpintero, 2016 lead to the same conclusion.

The Equivocality Problem opens up the following alternative. On the one hand, 'H-a-r-r-y-L-i-m-e' functions as a hanger for two names. This has to make it clear why the Inference would be unsound, but brings with it the need for making intelligible the link between the guy that acts in the story's plot and the fiction character created by Greene and Reed. This is Thomasson's stance. On the other hand, if there is only one proper name 'Harry Lime' which always makes the same semantic contribution, viz. regardless of the kind of context we look to, the Inference is sound. This is my choice. The obstacle I have to overcome is that of coupling the idea that the proper name 'Harry Lime' does not refer to anyone with the firm intuition that any reader/spectator of *The Third Man* may think of and talk about him.

2.4 Fiction Proper Names and the Causal-Historical Theory of Reference

In *Fiction and Metaphysics* Thomasson does not specifically consider the question of what fiction names refer to in the context of fiction.[19] (She repeatedly remarks that the discourse that belongs to that kind context must involve an operator of the form 'In the story *H*' or 'In the fiction *f*'.) Instead, she lends attention to the question of what the relationship is that fiction names hold to their *designata in the real context*. In this respect, Thomasson adds some new and interesting suggestions to the general debate. Her main thesis is that in the real context fiction names designate o refer to artifacts —([T3], in § 2). To the question of what supports the claim that they have such a kind of *designata* Thomasson provides an answer that substantially rests on a causal-historical theory of reference (Donnellan, 1972; Kripke, 1972; Evans, 1982, 1985, chap. 9). The core ingredient of that theory consists of two clauses. The first one says that a proper name gets linked to a person, an object, a place or whatever—an individual, in the logical sense of the word—through a baptismal ceremony carried out in more or less formal terms. According to the second one, once the ceremony has established the connection, the use of the name spreads along causal chains of communication that bring those individuals to bear upon other people's thoughts and speech acts. In that way, a name's user gains the capacity to refer to the individual on which the name was bestowed. "In general", Kripke says, "our reference depends not just on what we think ourselves, but on other people in the community, the history of how the name reached one, and things like that. It is by following such a history that one gets to the reference" (Kripke, 1972, p. 301).

Well, the problem that Thomasson must face up is that of explaining what relates a certain use of e. g. 'Harry Lime' to Harry Lime, the artifact. The answer cannot be

[19] I suspect that the reason why this is so is her remaining silent on the nature of imaginative acts. While in *Fiction and Metaphysics* she keeps herself far from fictionalism (see Thomasson, 1999, chap. 3), she opted for such a stance soon afterwards, though her commitment to fictionalism limited to textual and paratextual uses of fiction names. See Thomasson, 2003a, 2003b, 2013, 2015.

that there is a fictional character at the very beginning of the causal-historical chain, because, as [T3] states, that individual would be an abstract one, and nothing abstract keeps causal and historical relations to speakers that belong to some linguistic community. The metaphysical nature of the *designatum* makes it impossible for any baptizer to give it a name, thereby setting up a practice which other community members could join to. Thomasson introduces two novelties immediately after. He first modifies the communication chains so that they go back not to an individual—not to an abstract entity, in the fictional case, but to a either a text or another narrative medium that has a location in space-time. Texts and films are considered concrete individuals. The speaker thus refers to a fictional character by means of a fiction proper name provided that, as Thomasson points out, there is a causal-historical communication chain starting out at an author's vehicle in which the name somewhat appears.

Although there can be no direct pointing at a fictional character on the other side of the room, the textual foundation of the character serves as the means whereby a quasi-indexical reference to the character can be made by means of which that very fictional character can be baptizes by author or readers. Something counting as a baptismal ceremony can be performed by means of writing the words of the text, or (if the character is named later, for example by readers), it can remain unrecorded in the text (Thomasson, 1999, p. 47).

Thomasson's second novelty is to understand that in writing a text or shooting a film the author somewhat *creates* new characters, which very often involves naming them. In this way, "the paradigm case of naming involves recording that name in the text" (Thomasson, 1999, p. 48). The giving a character a name very often helps furnishing the character an identity of it own. As for the reader and the spectator, they refer to the character by pointing at him through a practice that Thomasson describes as *quasi-indexical*. They indirectly refer to the character by pointing at the text in which the character's name occurs. Therefore, in the first place a causal-historical reference chain is set up which goes back to either a text or a film script. Secondly, a name is grafted on a character by inserting the name in the text or in the script and articulating a system of cross-references that binds it. It is as if, as a result of so maneuvering, the speaker said either to himself or to some audience: 'By means of 'Harry Lime' I speak and think on the man who appears in *that* film', 'By means of 'Ofelia' I refer to the woman whom *that* drama displays as so-and-so'. We have here an act that not improperly might be thought of "as a kind of indexical reference" (Thomasson, 1999, p. 47).

This line of argument, subtle as it is, does not take into account the Equivocality Problem. It leaves in the dark the fact that fiction names that are used in the context of fiction hold an intimate relation to their uses in the real context. This relation straightly makes sense of readers and spectators' inheriting and an appropriating of the referential resources which the author employed. Instead of taking this insight seriously, Thomasson changes the *designatum*'s metaphysical category in moving from one kind of context to another. The critical movement is that of claiming that pointing at a text—a concrete entity—is enough to give rise to a link to an abstract particular. This shift from there being no referent to there being an abstract one is built into the Artifact Theory. 'Harry Lime' names nobody in the context of fiction, but what Thomasson takes to be the very same name designates something in the real context. A neat semantic difference maps on a metaphysical distinction, thus aking it impossible to dodge the Equivocality Problem. Thomasson disregards the problem, because she thinks she can accommodate two ideas: (i) that the same name occurs twice in the Inference's premises; and (ii) that their purported *designata* do not belong to the same ontological category. The causal-historical theory of reference hinders reconciling (i) with (ii). In raising these doubts, I would like not to be unfair to Thomasson's stand. Her strategy of turning to the causal-historical theory of reference to conclude that fiction names designate artifacts, though a sophisticated one, does not tally with my own views. I see the causal-historical theory as a trascendental condition of the plot's being articulated, better that as an empirical theory of fiction names semantics. In my opinion, her way of putting the theory to use overlooks that those names are *already* doing their work *in* the story, and that we gear ourselves to them in gaining access to its plot. Anyone who sympathizes with the theory's fundamental insight would have no reason to deny that *The Third Man*'s plot is so framed that Holly Martins refers to his former friend with the name 'Harry Lime' and that Major Calloway and Anna do the same. Of course, I am not claiming that those reference chains are real. They help framing the author's intimations, thereby making it possible to prop the plot up. The causal-historical theory is an important component of the fiction's hidden scaffolding, a trascendental condition for fiction unfolding.

2.5 After Anscombe's Footsteps

The alternative to the Artifact Theory I am going to recommend, the Intentional Object Theory, is imbued with the import of Elisabeth Anscombe's analysis of intentionality, as deployed in her classic paper "The Intentionality of Sensation" (Anscombe, 1965). Her enquiry deals with how sentences whose main verbs are intentional, i.e. 'to see', 'to hear', 'to think about', 'to intend', 'to say', as well as many others, are to be understood. To this list I will add 'to imagine' and speak of *intentional sentences* to bring those sentences into focus. It seems to me that Anscombe's way of handling them may be recruited to mold the basic insight that my alternative to Thomasson's views makes the most of. No doubt, the analysis of intentional sentences and the analysis of intentional acts are two poles apart. In spite of their being separate enterprises, I take Anscombe's results as supplying the Intentional Object Theory, if not a reliable handle in itself, a cue to get hold of one.[20]

[20] Crane, 2010, p. 91 underlines that the intentionality of a mental act does not reduce to the roles deployed by certain grammatical structures. He suggests this view to have found inspiration in a number of remarks Wittgenstein made in his *Philosophical Investigations*, and that Anscombe could have participated into such an anti-metaphysical attitude. (I gather that the remarks that Crane has in mind are Wittgenstein 11,953/2000, §§ 428–465. Whether Crane is right or not, my

Anscombe opposes two ways of getting the hang on the notion of a direct object when this notion is used to make the ins and outs of intentional sentences manifest. The first one is the modern way, which, according to her, is the one used by grammarians and linguists, and the old way. The second one, which is Anscombe's choice, has it that the direct object of sentence (5)

(5) John sent Mary a book

is the noun phrase 'a book', that is, a part of the sentence. On the old approach, in order to answer the question that asks what is (5)'s direct object, one must take an indirect route:

> However, the matter is not so easily settled. Of course, I do not want to oppose the practice of grammarians. But it is clear that the concept of a direct object—and hence the identification of the sentence-part now called the direct object—is learned somewhat as follows: the teacher takes a sentence, say «What did John sent Mary?» Getting the answer «A book» he says: «That's the direct object.» Now the question does not really suppose, and the pupil, if he goes along with the teacher, does not take it, that any particular people, of whom the sentence is true, are in question, and so we may say that when the teaching is successful the question is understood as equivalent to «What does the sentence 'John sent Mary a book' say John sent Mary» The grammatical concept of a direct object is acquired by one who answer any such question (Anscombe, 1965, p. 6).

The same strategy can be applied to make out (6)'s direct object:

(6) Reed imagined Harry Lime and Holly Martins talking to each other in a cabin of Prater's big wheel

The question could be: «Whom did Reed imagine talking to each other in a cabin of Prater's big wheel?» The answer would be: «Harry Lime and Holly Martins». Or perhaps, «Harry Lime» and «Holly Martins»: two answers, two direct objects. As Anscombe writes, nobody is to assume that the answer is true of two *specific* people—two guys within either the speaker's or the hearer's ken. Well, I suggest going a step further and adopting the concept of a direct object in Anscombe's old way as a model of the concept of intentional object. Thus, instead of asking after a sentence's direct object, we might try identifying the intentional object of Reed's imaginative acts, e. g. whom Reed imagined talking to each other in the Viennese Prater.

To articulate a view of imagination on the model of a grammatical treatment of intentional sentences is not out of place in this case, or so I gather. Anscombe finds the old way of handling intentional sentences liberating, for it avoids two alternatives stuffed with difficulties. The first one is the option of seeing a sentence's direct object as nothing save a part of a sentence. The second one is the alternative of conceiving direct objects as expressions that make a separate, autonomous semantic contribution. As Anscombe points out, (5)'s direct object could then be "any particular people of whom the whole sentence is true". The first alternative follows a nominalist course, whereas the second one fits in with a realist semantics. She

reliance on Anscombe's analysis is animated by the opposing attitude. My proposal chooses to think of Anscombe's remarks as shedding light on the nature of intentional acts.

rejects both views. On the option she chooses, there are direct objects *as far as* they figure as syntactic constituents in intentional sentences. Its is tempting to parody Frege and say that only in the context of a sentence is a linguistic expression a direct object. (I cannot resist the temptation.) In parallel to this analysis, judging Harry Lime and Holly Martins to be intentional objects discounts both their being nothing more than parts of sentences and their being separate, autonomous, entities.[21] On the option I choose, there are intentional objects *as far as* they participate as constituents in the contents, that is the thoughts, expressed by intentional sentences. Again, its is tempting to parody Frege and say that only in the context of an imagination act is does a thought have an intentional object as a constituent. Thomasson's Artifact Theory agrees with the realistic alternative. As for the Intentional Object Theory, it follows Anscombe's footsteps: there are fiction characters in so far as they figure in imagination acts.

2.6 Fiction Characters as Intentional Objects

What does it mean that Graham Greene created a character to whom he gave the name 'Harry Lime'? It means that Greene imagined Lime as well as the plot in which the latter plays a significant part. Harry Lime only exists in this fiction and 'Harry Lime' is the name of an imaginary person. For anything to be an imaginary individual is for it to exist as depending on someone or something else. An imaginary individual is not a metaphysically autonomous individual. Harry Lime is an intentionally dependent individual[22]: it does not exist independently of those acts by means of which it is imagined. Being a fictional character, it exists insofar as it features as an object in an intentional act, e. g. an imagination act. These claims state what Thomasson calls the Intentional Act Theory. Fiction characters are the individuals that appear in fiction. We use fiction names to talk about and refer to intentional objects, objects which, unless they be real, only exist as objects of imagination acts.[23]

Thomasson discusses the Intentional Act Theory when she considers Sartre's defense of it in *L'imaginaire* (Sartre, 1952). She admits the theory to be close in spirit to her own alternative—the Artifact Theory. However, she is at variance with Sartre, since she rejects that an imaginary individual is "an entity created in an imaginative act of consciousness and *that exists only as long as it is being imagined*" (Thomasson, 1999, p. 21 f. My italics. J. J. A.). She justifies her opposition to

[21] It is important, to grasp the scope of the Intentional Object Theory, to take heed of the differences between the concepts of entity and object. On this theory, fiction characters, i.e. what fictiodesignators refer to, are not entities; they are objects of intentional acts. See Crane, 2001, 2010, §§ 4.1–4.2, where the difference is clearly traced out.

[22] Husserl, 1913/1952: §§ 87–96; Ingarden, 1968/2005: § 8; Thomasson, 1999, pp. 88ff.; Crane, 2001; Sainsbury, 2010.

[23] This view is argued for in Crane, 2010.

Sartre's views with two arguments. The first argument's main premise contradicts the claim that objects only exist as long as they are being imagined. She adduces that such a view "runs counter to our usual practices in treating Holmes, Hamlet, and the rest as enduring through those periods of time in which no one is imagining them" (Thomasson, 1999, p. 22). The second argument questions Sartre's thought that not only an author's imaginative acts create a fictional character, but also that his readers recreate the same character *afresh*. Thomasson then emphasizes the difficulty of explaining, "how we can legitimately say that the two or more readers are each reading about or experiencing one and the same fictional character" (Thomasson, 1999, p. 22).

Thomasson's first argument against the Intentional Act Theory relies on the assumption that when we are not imagining Harry Lime, when, for instance, we are neither reading Greene's novel nor watching Reed's film, we somewhat count on Harry Lime's existence. Since she excludes his existence in not imagining him, she cannot count on his existence *then*, the reason being that we are *not* going through any intentional act that contains Harry Lime as an object. The difficulty lies in making sense of the idea that there is some Harry Lime whenever we do not think of him. How can we count on him, save by imagining him? Well, in spite of its apparent force, Thomasson's response is not a conclusive one. She seems to think that an insuperable gap insulates Greene's imaginative acts from the text in which those abilities are fixed. I concede her that imaginative abilities may be distinguished from the literary and cinematographic works which serve as their vehicles. In spite of this, it is far from obvious that any author's imaginative acts do not remain in his or her works. (We do judge testaments of expressions of people's last will.) Thomasson concludes that they do not, since she identifies imagination with imagination *acts*, that is, with episodes or occurrences which involve what agent imagined—an intentional object—at a time. Once this step is taken, texts and films, which make it possible for those acts to become stabilized, are left out of the setting. This conclusion is too controversial. Thomasson should agree to the remark that imagination acts remain alive only if the very material works are neither destroyed nor forgotten. Texts and films can be so conceived that they not only register the authors' intentional acts, but also potentially contain their readers' and spectators' imaginative recreations. Literary and cinematographic works do not actually contain the relevant acts, but they potentially do it. Words, sentences, images, recordings, etc. are frozen acts of imagination. Intentional objects may not be present when there is no act of imagining so-and-so itself, provided that 'act' is understood in an episodic sense, and still be present as objects of intentional *potential* acts. Thomasson's claim that "a fictional character is created by an author who constructs sentences about it, but it is maintained in its existence thereafter not by the imagination of individuals, but by the words and sentences themselves" (Thomasson, 1999: 22) cannot be literally endorsed. That the words and sentences themselves can be detached from the intentional acts that brought about them and still fulfill their function is a mirage.

Intentionalists are of the opinion that a literary work's words and sentences hold an intimate relationship to the intentional objects set up in the imagination acts that gave rise to those works. Therefore, both the Artifact Theory and the Intentional Object Theory understand fiction characters as metaphysically dependent entities. The theories differ as for the dependency basis. An artifactualist takes this basis to be the texts and films; intentionalists focus on imagination acts, either occurring or potential. Thesis [T4] (see § 2) is the crossroad where both approaches part ways. The Artifact Theory incorporates it; the Intentional Object Theory rejects it. Thomasson thinks Harry Lime, the cultural artifact, to possess a higher ontological autonomy than Harry Lime, the intentional object. The Intentional Object Theory has it that the only autonomous thing is the imagination act, its intentional object being a non-separable constituent of the act. Putting together two distinct realities, which antecedently exist, namely the act and the object, does not form intentional acts. The intentional object comes out in the intentional act's constitution.[24]

Thomasson's second argument against the Intentional Act Theory centers on a crucial problem in the philosophy of fiction. The problem, she thinks, fatally affects Sartre's views on imagination. On the one hand, why does Sartre would say that two spectators of Reed's *The Third Man* recreate its plot *afresh*? On the other hand, how could it be defended that they have watched the same film and shared the same intentional object Harry Lime, if each of them has recreated the plot from the very beginning? I will not comment on the first question in these pages, since it addresses so complex a subject that a separate essay is needed to do justice to it.[25] As for the second one, absorbing as it is, it does not have to get intentionalists into a muddle. Thomasson would face the question by pointing out that the film includes sequences enough, both quantitatively and qualitatively, to create an abstract entity whose existence would not depend on the spectators' imaginative acts. Therefore, two different showings of the same film would put on display the same characters, i.e. the same abstract artifacts, to different spectators. Some of their thoughts and emotional reactions can follow their own specific routes and be addressed at the same characters. (The same is true of two tourists who gaze at the Empire State Building, one at

[24] It is not easy to be more specific about the kind of dependence that is respectively built into intentional objects and artifacts. A promising line resorts to the concepts of autonomy and separability. On it, intentional objects are neither autonomous nor existentially separable from the acts in which they figure. Artifacts, on the other hand, are not autonomous *vis-à-vis* the texts on which their existence depend (Ingarden, 1964, pp. 79ff.), but are separable from them (Ingarden, 1964, pp. 115ff.). The artifact Harry Lime is separable from Greene's novel, because destroying all its copies would not mean the end of the artifact's existence. If the plot remains in someone's memories, Harry Lime would still be out there. The key difference between the Intentional Object Theory and the Artifact Theory would lie in how strong are their respective links. Another way of accounting for the the different metaphysical status of artifacts and intentional objects is by resorting to Thomasson's distinction between constant dependency and historical dependency. Intentional objects would constantly depend on imagination acts and, as a consequence of it, historically depend on them as well. As for artifacts, they would historically depend on texts and films, but their dependency would not be constant. This conjecture needs more elaboration. See Thomasson, 1999, chap. 2.

[25] Here we confront the Intentional Identity Problem, how to explain that one fiction character can be the same intentional object of two people's intentional acts. On the nature of intentional identity, see Geach, 1972, § 4.4. For a formal treatment different from Geach's, see Saarinen, 1979a, 1979b.

its foot and the other on a helicopter at 800 meters above street level.)[26] Thus, there is a neat reply to Thomasson's second objection, namely, that the fact that two or more spectators have to recreate the same plot afresh does not imply that their results wildly differ from each other. Both recreations share the same materials, and unless these materials are scarce and clumsily handled, heading the plot into indeterminate margins, the character's identity is not per force in question. Of course, the spectators' cultural background can make a significant difference. Therefore, Thomasson is right depending on who the two spectators are. However, there is no principled reason to deny that in getting acquainted with the plot, readers and spectators often become in rapport with the author's intentional objects. True, this does not mean that disagreement might arise concerning one or another aspect of any character's deeds and features. I can believe that Harry Lime is fooling Anna while you think it otherwise. Or you might ponder some doubts over what feelings Holly Martins harbor towards Harry when Martins finally shoots Lime down in the Viennese sewer net. The fact that we have a grip on other people's intentional objects underwrites their objectivity, because it gives readers, spectators and critics guarantees that not anything goes in approaching them. A variety of stances may be adopted towards intentional objects without having to conclude that there is no fact of the matter concerning many questions that can be asked about them. Leaving aside the author's competence, arguments can very often be given pro and con that such-and-such character is so-and-so. The specific features of the work can make easier for readers to concur with each other, but they can also dissolve a character's identity. All depends on how complete the work's design and deployment are. This can be conceded and at the same time reject the thought that common recreation is exceptional.

2.7 Intentional Objects á-la-Goodman

How can the Theory of Intentional Object help to assess the Inference? I will set it forth by exploiting an idea of Nelson Goodman's. Goodman suggested that predicates like 'picture-of-an-unicorn' can be used to classify pictures (Goodman, 1976: § I.5; 1984: § III.3). A picture of a unicorn does not commit anyone to the existence of unicorns. Neither the concept of a unicorn nor the predicate 'picture of-an-unicorn' do commit anyone to the existence of unicorns. That there are unicorns does not follow from the true thought that that picture—pointing at one—represents a unicorn. Neither does it follow from a true statement to the effect that *that*—pointing at one—is a picture of a unicorn. The predicate '(is an) unicorn' does not make

[26] The reader can check that Thomasson does not provide either this argument or something close to it in facing what she calls Sartre's problem. She repeats her argument against the claim that intentional objects only exist when we are imagining then, but this is not the question she ought to treat. The question that matters is why two different readers can address the same intentional object. Cf. Thomasson, 1999, p. 23.

a contribution of its own. It is my view that a fiction name like 'Harry Lime' does not force readers and spectators to commit themselves to the existence of Harry Lime either. Let us go into the details.

A predicate like (7) admits of two different analyses: (7.1) and (7.2)

(7) a picture of Paris
(7.1) a picture of (Paris)
(7.2) a-picture-of-Paris

The difference between (7.1) and (7.2) is that in (7.1) 'Paris' makes a contribution of its own, viz. that of involving the very city of Paris in the role that the whole expression plays. In (7.2), however, the predicate is not taken as a complex whose constituents are structurally organized. The predicate as a whole is what (7.2) is meant to be true of. It is so understood that the predicate is true if it is a picture of Paris and false otherwise. Taking advantage of Goodman's suggestion, we can understand (8) either as (8.1) or as (8.2):

(8) a novel on Harry Lime
(8.1) a-novel-on (Harry Lime)
(8.2) a-novel-on-Harry-Lime

(8.1) treats the fiction name 'Harry Lime' as a referring expression in its own right, that is, as a tool whose use presupposes that there is someone represented by it—a fiction creature. (8.2) does not go beyond helping us to classify novels. Any commitment to the existence of fiction characters is cancelled. Goodman's proposals make it possible to integrate proper names' occurrences into expressions that belong to other logical categories. One possibility is that of integrating them into predicates, as (8.2) illustrates. How could we deal with (9)?

(9) The police are on Harry Lime's trail

Here is a first answer:

(9.1) Is-on-___'s-trail (the police)

In (9.1) the name 'Harry Lime' has been merged into the predicate, while allowing the designator 'the police' to carry out a referential contribution of its own. This way of interpreting (9) is unsatisfactory because, although (9) is used to say something of *The Third Man*'s plot, what it conveys is false. No real police force is after Harry Lime, a non-existent criminal. Thus, instead of (9.1) a better analysis of (9) would be either (9.2) or (9.3):

(9.2) The-police-is-on-Harry-Lime's-trail
(9.3) λp [p] (the-police-are-on-Harry-Lime's-trial]

Whereas (9.1) would be true in case that a real police force was tracking Harry Lime, (9.2) is not to meet such a condition. The disadvantage of choosing either (9.2) or (9.3) is that they hardly amount to any analysis at all. Goodman's clever trick seems to crack. But wait! It nicely works when it comes to statements used in the real context, as in (10):

(10) Harry Lime was created by Graham Greene

Here we can merge again the name 'Harry Lime' into the predicate, as (10.1), (10.2) and (10.3) show:

(1o.1) Harry-Lime-was-created-by (Graham Greene)
(10.2) The property of *having-created-Harry-Lime is* possessed/exemplified by Graham Greene
(10.3) λp [p] (having-created-Harry-Lime) (Graham Greene)

A similar analysis leads us from (11) to (11.1):

(11) Many British policemen admire Sherlock Holmes
(11.1) (Many British policemen x) (Admire-Sherlock-Holmes, x)

It can be seen that this analysis works reasonably well with statements used in the real context that contain fiction names. However, it cracks when it comes to fiction names used in the context of fiction. (10.1) and (10.2) are satisfactory renderings of (10), whereas (9.1) does not improve on (9), for in (9.1) 'the police' involves not a real police force, but *The Third Man*'s police, the imaginary police who is on Harry Lime's trail. What we need is a uniform way of dealing with fiction names no matter where they occur, whether in statements used in the context of fiction or in statements used in the real context. The analysis should not come into conflict with Lewis' claim that Harry Lime, Sherlock Holmes, Hamlet, Don Quixote, etc. are persons. By getting round the conflict, we dodge attributing to them uncommon, exotic features. The analysis should also avoid treating fiction names as expressions that make a separate ontic contribution, e.g. that of involving a fiction entity. In other words, to give up thinking of those expressions as having got the function of representing somebody or something. Put it in the old, classical way, to give up consider them as cathegorematic expressions.[27]

It is important to add, to get things clear hereafter, that this proposal goes along a territory in which two different forces exert their powers. One of them is the pressure of having to display the *meaning* of those statements—an exercise in semantic analysis. The other one is the pressure of having to bring the ontic commitments implicit in using names of fictional characters to light. (I will use the term 'onto-semantic' to make the two-targeted character of the task clear.) Therefore, the aim is to set forth the truth-conditions of those statements while being faithful both to their meaning and to the intentionalist's demands. In this spirit, a first step of the analysis of (2) is shown in (2.2):

(2) Harry Lime traffics in diluted penicillin
(2.2) The-Harry-Lime-of-(*The Third Man*) traffics in diluted penicillin

[27] Everybody knows that the paradigmatic application of this strategy in Analytic Philosophy is Russell's theory of descriptions. The Goodmanian maneuver is close in spirit to Russell's fundamental principle of the theory of denotation. See Russell, 1905/1994.

Generalizing, if *N* is a fictional character's name that appears in a literary or cinematographic work *k*, (12) becomes (12.1):

(12) *N* is *P*
(12.1) the-*N*-of-(*k*) is *P*

When *k* is Shakespeare's *Hamlet* and *N* is 'Ofelia', (13.1) gives rise to (13.2).

(12.2) The-Ofelia-of-(*Hamlet*) is *P*

In (13.2) 'the-Ofelia-of-(*Hamlet*)' is a singular term, that is, a designator by means of which we talk about a character of Shakespeare's *Hamlet* without allowing the proper name 'Ofelia' to develop its own onto-semantic function. Thus, the name 'Ofelia' is treated as a syncategorematic expression. It has no referent. This idea provides an outline of how to throw light on the use of fiction names both in the context of fiction and in the real context. Thus, (1), (2) and (9) respectively become (1.1), (2.s) and (9.3):

(1.1) The-Harry-Lime-of- (*The Third Man*) is the character created by Graham Greene in *The Third Man* whose role was played by Orson Welles in Carol Reed's film
(2.2) The-Harry-Lime-of-(*The Third Man*) traffics in diluted penicillin
(9.4) The-police-of-(*The Third Man*) is after the-Harry-Lime-of (*The Third Man*)

It must be noticed that expressions of the form 'the-*N*-of' are functors, i.e. functional expressions. They give rise to a designator when combined with one designator. I what follows I will call designators of the form 'the-*N*-of-(*k*)' *fictiodesignators*. In the examples considered here, they give rise each to a designator when combined with a proper name. As a consequence, this analysis needs two designators to materialize: the argument of the functorial expression and the result of combining this with a proper name. From an onto-semantic point of view fiction names ('Harry Lime', 'Ofelia', etc.) are neither names nor fictiodesignators.[28]

It is natural to wonder why to reject that fiction names do not make an onto-semantic contribution of themselves, while approving that fictiodesignators do. That is, why to reject that 'Harry Lime' refers to a fictional character and endorse that 'the-Harry-Lime- of -*The Third Man*)' designates something. The reason is that a literary work is an existent entity, and the fictiodesignator makes it clear that the

[28]The analysis I am developing is close in spirit to Geach's insight that expressions like 'Harry Lime' are *cuasi-names*, that is, expressions used without being committed—either by violating a condition or by ignoring a presupposition—to an existing entity. For Geach, it reduces to the condition that cuasi-names "are not available for use as a subject of predication" (Geach, 1972, p. 162), that is, they cannot occur in positions that sentences reserve to subjects of predication. The difficulty that Geach's suggestion brings up is that of giving form to the requirement of preventing cuasi-names from taking up those positions. It seems to me that the Goodmanian maneuver presented above is a way of achieving it. (There are other strategies that have the same target, but I will not take time over them here. See Predelli, 2002; Reimer, 2005; Voltolini, 2006.) Maybe Geach wanted to honor the condition that name fictions are names, because that is what grammar calls for. It is obvious that the Goodmanian maneuver is not in line with that condition.

fictional character does exist as far as this character features in the story's plot. There is no independent way of getting at it save through immersing ourselves in a novel or a film. This point is a metaphysical one: a fictional character depends for its existence on the work in which it appears.

The analysis of statements like (9) and (11) is as expected. Depending on whether we count on the property of being chased by the police or on the property of being chased, we respectively get (9.5) and (9.6):

(9.5) The-police-is-on (the-Harry-Lime-of (*The Third Man*))'s trail
(9.6) On- —'s-trail (the-the-police-of-(*The Third Man*)) (The-Harry-Lime-of-(*The Third Man*))

(9.5) is a natural rendition of (9), in spite of presupposing that the properties attributed to the fiction character are admissible in the real context as well. (Science fiction stories bring exceptions to such a rule to bear.) It is important to realize that it does not contain 'the police' as a singular term. In contrast to (9.5), (9.6) must subject the expression 'the police' to the kind of treatment we recommend for fiction names on pain of attributing (9) a meaning it does not possess. With regard to (11), it would be natural to prefer (11.2) to (11.1):

(11.2) (Many British policemen *x*) (Admire (*x*, the-Sherlock-Holmes-of-(A *Study in Scarlet*))

The same analytical pattern smoothly applies to those statements in which both fiction names and real people's names co-occur. (14) is an example of that type of case and (14.1) an analysis for it::

(14) Holly Martins meets Harry Lime in Vienna
(14.1) Meet-in (the-Holly-Martins-of-(*The Third Man*), the Harry-Lime-of-(*The Third Man*), Vienna)[29]

(14.1) treats the proper name 'Vienna' as the name of the real Vienna, i.e. that 'Vienna = the-Vienna-of-(*The Third Man*)' is true.

Let us now approach the Inference from the Goodmanian viewpoint outlined so far. (1) is rendered as (1.1) and (2) as (2.1). (1.1) is transparent: it says that the fictional character Harry Lime was created by Graham Greene. (2.2) furnishes the thought that *The Third Man*'s Harry Lime a linguistic format. In other words, (2.1) is true if, and only if, the Harry Lime whose deeds *The Third Man* tells traffics in diluted penicillin. That character is not the abstract entity Thomasson asks us to welcome, but the concrete guy who Lewis would believe in. Once again: the proper name 'Harry Lime' does not make any contribution to these truth-conditions of itself. In the Goodmanian analytical framework I have sketched, 'Harry Lime' is not the linguistic vicar of an intentional object. The only intentional object that the analysis brings to bear is the-Harry-Lime-of-*(The Third Man)*. In other words: the intentional object of Greene's imaginative acts is something which does not exist

[29] To avoid an unnecessary prolixity, I put Davidson-like analysis of event sentences aside.

independently of the work in which its name occurs. 'Harry Lime' does not stand for a fictional character; the fictiodesignator 'the-Harry-Lime-of-(*The Third Man*)' designates it. Both (1.1) and (2.2) are accountable to the principle that fictional characters are intentional objects. It turns out, then, that the Inference is sound, because it is an inference of the form: 'If $t_1 = t_2$, and $\alpha(t_1)$, then $\alpha(t_2)$'.

This result may be contested. It can be replied that by rendering (2) as (2.2) one has taken a decision without bear all the relevant alternatives in mind. The fact is that one of (2)'s uses has been left aside, namely the one we find in *The Third Man* when e. g. Major Calloway informs Martin Hollis that his former friend, Harry Lime, not only is alive, but also that he is trafficking in diluted penicillin. This remark points out that, since the sentence (2) can be used to make two different statements, the analysis provided above is definitively a *non-sequitur*. In setting (2) on the context of fiction, namely the context in which Major Calloway acts, a statement is obtained that cuts the course from the Inference's premise (1) to its conclusion (3). Therefore, the Inference is sound on one of the uses of 'Harry Lime' but false on the other. I have set forth why intentional objects are the answer to the question about fictional names' onto-semantic contributions in real contexts. However, the role of fiction names in the context of fiction has been out of my sight. What do they contribute to the truth-conditions of the sentences in which figure in the sort of context? My answer is that even in the context of fiction (2) is to be rendered as (2.2). It is not necessary to come off the way already open. The fictiodesignator 'the-Harry-Lime-of-(*The Third Man*)' deploys a referential function: it purports to designate a fictional character, although *in the context of fiction* it has no referent. Let us see why.

When happily used, fictiodesignators designate intentional objects. This means that they designate, and they purport to designate, individuals *as* imagined by an author. In contrast to what is the case in the real context, Major Calloway does not intend to pick up one individual among the denizens of Greene's imaginary world, because Major Calloway does not use the name 'Harry Lime'. He is imagined to do that. Greene is the name's user, and the context in which he uses the name is the context of fiction. This makes a significant difference, for the use of fictiodesignators, just like the use of functional expressions in general, presupposes that there is something to which one can refer with their help. In the case in hand, the presupposition is not satisfied. Greene *imagines* a plot, and Harry Lime is part of it. Greene knows that Harry Lime does not exist when he has Major Calloway informing Holly Martins. The use of 'the-Harry-Lime-of-(*The Third Man*)' in the context of fiction does not presuppose that Harry Lime exists. Therefore, the fact that such a presupposition is not satisfied explains why the fictiodesignator makes no contribution to (2)'s truth conditions.

2.8 ERE, OAS and the Principle of Equivocality

If 'the-*N*-of-(*k*)' is a monadic functor and *O* a proper name that can take up the place of its variable *k*, then 'the-*N*-of-(*O*)' is a designator, viz. a fictiodesignator, which can be used to refer to an individual. What kind of individual? The answer has been anticipated: fictiodesignators refer to fiction characters. The Intentional Object Theory has it that fiction characters are intentional objects, objects that, save they independently exist, exist as far as they figure in a fiction work. In other words, they do not exist independently of their works. The Goodmanian strategy tried out in the previous section (§ 7) is faithful to such a predicament. Statements that include fiction names, as well as thoughts which are about fictional characters, when opened up, reveal that any fiction character is logical and conceptually linked to a fiction work. The fictiodesignator 'The-Harry-Lime-of-(*The Third Man*)' *refers to* the character Harry Lime *insofar as it is a character in* Greene's novel (and Reed's film). The dyadic relative term '__ refers to…' has been replaced by a triadic relative term: '__ refers to … insofar as it is a character in --'.

To what extent does this approach to fictional characters differ from Thomasson's? The principle *ERE* is an appropriate point of reference to get the comparison to unfold. On letting the Intentional Object Theory and the Goodmanian treatment of fiction names work in tandem, three claims come to the fore:

(Int1) In the context of fiction, fictiodesignators have no referent.
(Int2) In the real context, fictiodesignators refer to intentional objects.
(Int3) Fictiodesignators meet *ERE* both in the context of fiction and in the real context.

On the Artifact Theory's part the following three counter-claims are offered:

(Art1) In the context of fiction, fiction names have no referent.
(Art2) In the real context, fiction names refer to artifacts.
(Art4) Fiction names meet *ERE* both in the context of fiction and in the real context.

It is plain that both series of claims run parallel to a significant extent. The main difference has to do with whether fiction proper names are thought of as cathegorematic expressions, that is, as making a contribution of their own to the truth-conditions of those sentences in which they occur. According to the Intentional Object Theory fiction proper names do not so function. On the Artifact Theory, they introduce an abstract entity into those conditions when used in the real context and do not make any contribution to them when used in the context of fiction. Anyway, the principle that a proper name '*N*' refers to an individual *x* if, and only if, *N* exists is safe from counter-examples. In the context of fiction, since the non-existence of any cultural entity *N* implies that '*N*', the name that would be paired to it, does not refer; and since there being nothing to which '*N*' refers to, *N* does not exist. In the real context, since the fact that '*N*' refers to a fictional character implies its existence; and since the existence of the artifact *N* implies, according to the theory, which the name '*N*' is there to name it. The Intentional Object Theory accords with the principle *ERE,* but it looks in a different direction. In the real context a fictiodesignator 'the-*N*-of-(*k*)' is used to refer to an intentional object. Intentional objects exist as constituents of imagination acts whose existence is taken for granted. In addition to it, as soon as an author imagines a character to take part in a story's plot *k*, within which it is called '*N*', the conditions have been arranged for that intentional object to be referred to by the fictiodesignator 'the-*N*-of-(*k*)'. When it comes to the context of fiction, the biconditional "*N*' refers if, and only if, *N* exists' holds because both conditions are trivially true. From left to right, because '*N*' does not refer to anything in the context of fiction; from right to left, because in that kind of context its is assumed that *N* does not exist.

It is true that the principle *ERE* sets margins within which the Intentional Object Theory and the Artifact Theory define their own territory. However, the principle forces some distance between them. The reason is that *ERE* bind us to think of fictional characters' existence in semantical terms. To be specific, the question is, what sort of entity is referred to by 'Harry Lime', in order for a statement like (1) or (2) be true, provided that a universe of discourse is somewhat fixed? Questions as these fit the orthodox approach in model-theoretic semantics—what does a designator *N* has to refer to, and a predicate *P* be true of, in order for the sentence *N is P* be true, provided that a universe of discourse is somewhat fixed? This is the key question, and (Art2) and (Art3) are the Theory of Artifact's answer to it. The theory's main movement consists of arguing that a cultural artifact, an abstract entity, is to be postulated to account for (1)´s truth-conditions. On the other hand, the Intentional Object Theory holds that the fiction discourse is not a subject to be dealt with in model-theoretical terms. Intentionalists thereby avoid the model-theoretic tour. Intentionalists hold that semantics should be left momentarily aside and give instead priority to setting the focus on imagination acts. This change of emphasis is not meant to completely disregard any semantic treatment, but to postpone it until we have derived benefits from an intentional act analysis. The main benefit so obtained is that fictional characters exist as *objects* of those acts. (See note 21.) To hold that Harry Lime is an object of Greene's imagination acts is to hold that by grasping them we become entitled to answer questions like the following: Who did Greene imagine to have pretended his own death? What event did Reed imagine to take place in a cabin of Prater's big wheel in Vienna? With whom did Reed imagine that Harry Lime hold a sentimental relation?

Anscombe lesson, expounded above (in § 5) in the formal mode of speech, suits the Goodmanian treatment of fiction names reasonably well. Hence, on the Intentional Object Theory those names are not in conflict with *ERE*—as (Int3) sets down. If there remains still one doubt, it concerns the sort of *designata* that fictiodesignators have, since room has been left to debate whether intentional objects do exist. I have claimed that there are fiction characters exclusively as objects of imagination acts, objects whose names figure in their respective stories. (Int1) and (Int2) put on record my commitment to this thesis. Thomasson may find here the opportunity to disagree. She could concede that intentional objects exist and then argue that the sort of existence they have is not separable enough from the imagination acts in which they are embedded. The criticism would be that such a connection is unduly strong, and that a more flexible sort of bond—like the one provided by a causal-historical theory of reference—would better account for fiction names' uses. On sticking to this argument, *ERE* would not be honored and I should give up (Int3). This is not the end of the story, however, because the alternative of reducing ontological pressures, by adducing that intentional objects are available to public scrutiny, is not without any merit. The counter-argument would now reiterate that the Intentional Object Theory obeys *ERE*, because intentional objects' intentional identity is beyond doubt, and this speaks in favor of their existence. No matter what alternative we choose, a real asymmetry arises here, because Thomasson's artifacts do not keep with respect to their vehicles, viz. texts, films, recordings or memories, the kind of relation that intentional objects keep to their constitutive acts.

What about *DOS*, the principle that has ontology depend on semantics? Do the Intentional Object Theory and the Artifact Theory obey this principle? In § 4 I put forward Thomasson's account of how fiction names designate existent individuals in the real context—specifically, of how they refer to abstract cultural entities—and its reliance on a causal-historical theory of reference. I objected to that view that in appealing to it Thomasson heads for a clash against the Equivocality Problem. As for the Equivocality Problem, intentionalists and artifactualists part ways. In § 7 I have taken advantage of a Goodmanian trick, which take fiction names to be syncategorematic expressions. The aim was to endow the Intentional Object Theory with a logico-linguistic frame that reveals the ontic commitments assumed by the use of fiction names. One benefit of so proceeding is that of allowing intentionalists to avoid facing the Equivocality Problem. They are committed to holding that fiction names' *designata* belong to one and the same ontological category no matter what context of use is considered. Artifactualists and intentionalists differ about *DOS*. Whereas Thomasson's follows the principle, intentionalists come into conflict with it. Artifactualists seem not realize that the conjunction of (Art1) and (Art2) pushed them to admit that 'Harry Lime' is a two-symbol, equivocal, expression. Intentionalists are aware of that and look for a strategy that allows them to bypass the Equivocality Problem. To sum up this discussion, to (Int1) - (Int3) and (Art1) - (Art3) we must add (Int4) and (Art4):

(Int4) Fictiodesignators do not fit *DOS*.
(Art4) Fiction names fit *DOS*.

Thomasson's reply to (Int4) is that the Intentional Object Theory's denial that 'Harry Lime', 'Ofelia' and the rest are proper names is revisionist. Thomasson is right. She seems to abide by the so-called Grammatical Constraint, according to which the differences between syntactic and semantic structures should be minimized. This constraint is violated by (Int4), which turns what look like names from a syntactic point of view into semantically non-separated pats of functorial expressions. The difference between syntax and semantic is undeniable. On Thomasson's view, fiction names are proper names, from the viewpoint of both syntax and semantics. It follows that semantics, that is, the theory of reference, should have priority over the analysis of intentional acts.[30] Of course, the Intentional Object Theory disagrees. In my view, the way out from this impasse lies in realizing that there no insurmountable barrier between semantics and metaphysics. At most, there are conflicts among semantic theories themselves, on the one hand, and among metaphysical analysis themselves, on the other hand.

2.9 Semantics *vs.* Metaphysics?

I will finish this contribution by briefly showing how to transform the Goodmanian analysis of fiction names sketched above into a semantic theory that accords with *DOS*. Since I only want to cast doubts on the revisionism charge, that would call the intentional approach into question, I will not pause to present any evidence for that theory.

On the indexical theory of proper names (Recanati, 1993, § I.4.8), a proper name N is an expression whose referent, x, depends on context. The context that matters is a convention that prevails in the community in virtue of which N refers to x. Thus, as a proper name 'Everest' refers to an individual x, viz. the well-known mountain, relatively to a context that links the name to the mountain. This theory of reference makes the role of context explicit by adding an index i to the proper name N. 'Everest$_i$' refers to the Everest, provided that i is the convention that links the expression 'Everest' to the mountain. Generalizing: a proper name N_i refers to x, if i links N to x. The Spanish proper name 'Cristóbal Colón$_i$' (i.e. the English proper name 'Christopher Columbus') might in principle refer (in Spanish) not only to the admiral who arrived at La Española (Dominic Republic) in October 1492, but to another Spanish admiral, Cristóbal Colón de Carvajal y Maroto, who lived from 1925 to 1986. 'Cristóbal Colón$_i$' refers to the former, if i links the expression to him; it refers to the latter, if i links the expression to Columbus' descendant. Whom the name refers to (in Spanish) depends on the context. One context was the convention in force in 1492; the other one since 1925. Two conventions, but only one name. The second admiral, it may be naturally said, was baptized with the same name as his distant ancestor. Thus, we may identify the proper name 'Cristóbal Colon' with the expression 'Cristóbal Colón$_i$'. The index i takes conventions as values.

The same rule applies to fiction names, assuming that they are to be processed by a semantic theory of natural language. The already familiar proper name 'Harry

[30] Anyway, the alternative of being or not being a revisionist is not a new one. Geach did not conceal his stand as for this matter, arguing that metaphysics do not have to run parallel to grammatical analysis. An expression like 'Harry Lime' might not be considered as a designator in seeing to metaphysical needs, in spite of having to give up pursuing grammar theory's aims. The metaphysicians' objectives and methods do not have to be dictated by the grammarian. (And the other way around.) Geach took fiction names to be and proper names from cuasi-names from a logico-semantic viewpoint and proper names from a syntactic viewpoint. Of course, this position violates de Grammatical Constraint.

Lime' is hereafter represented as 'Harry Lime$_i$'. One implicit convention would be in force in Greene and Reed's plots in virtue of which 'Harry Lime' would refer to Holly Martins' friend and "Baron" Kurz and Popescu's partner in crime. Strictly speaking, neither Greene nor Reed openly states the convention. However, they set up the conditions that the convention to be activated as an ingredient of the story's scaffolding. According to the convention that governs the use of 'Harry Lime$_i$' in *The Third Man*, the name's referent is no other than the referent of 'Harry Lime$_{\textit{The Third Man}}$', i.e. the imaginary man whose dark deeds the story tells. I will use the expression 'Harry Lime$_{\textit{The Third Man}}$' to represent the proper name as used in accordance with the convention implicit in *The Third Man*. Moreover, this line of analysis can be enriched by putting on record who introduced the convention. New representational possibilities are thus created:

(17) Harry Lime$_{(\textit{The Third Man}, \text{Greene})}$ = Harry Lime$_{(\textit{The Third Man}, \text{Reed})}$

This statements says that the fictional character referred to by 'Harry Lime$_i$' in the context created in Greene's novel is the same character as the one referred to by that very name in the context of Reed's film. If Reed was faithful to the novel—and he was, because Greene took part in giving form to the film's script—we may infer that (17) is true. And if we agree that the Philip Marlowe character in *The Little Sister* is the Philip Marlowe who figures in *The Long Goodbye*, then we will conclude that the statement (18) is equally true:

(18) Philip Marlowe$_{(\textit{The Little Sister}, \text{Raymond Chandler})}$ = Philip Marlowe$_{(\textit{The Long Goodbye}, \text{Raymond Chandler})}$

By means of (18) we can assert that the fiction character referred to by 'Philip Marlowe$_i$' in virtue of the convention in force in *The Little Sister* is the same character as the one referred to in virtue of the convention activated in *The Long Goodbye*.

Approached from the indexical theory of reference, the Inference turns out to be sound:

(19) {Harry Lime$_{(\textit{The Third Man}, \text{Greene})}$ is the character created by Graham Greene in *The Third Man* whose role was played by Orson Welles in Carol Reed's film; Harry Lime$_{(\textit{The Third Man}, \text{Greene})}$ traffics in adulterated penicillin} ⊨ The character created by Graham Greene in *The Third Man* whose role was played by Orson Welles in Carol Reed's film traffics in diluted penicillin

The most delicate aspect of (19) is the assumption that the proper name 'Harry Lime$_i$' is used in the two premises in accordance with the same convention. I think that this is the case. It sounds very odd to hold that there are two different conventions involved: one governing the use of the name to pick up one *designatum* in order to predicate of him that he traffics in adulterated penicillin; and another backing the reference to quite a different kind of being. The meaning of (2) relies on a convention that links 'Harry Lime$_i$' to a fiction character. That a second convention is needed to tell its etiology is, to say the least, striking.

On the other hand, a certain affinity can be perceived between justifying the Inference from a semantic perspective and justifying it from the intentionalist stance. In other words, between the analysis furnished by the indexical theory of reference and the one enabled by the Goodmanian maneuver. Both exercises lead to conclude that the Inference is sound. This is hardly surprising, since there exists a way of coordinating both perspectives. One can see what the coordination key is by looking at the example which I have been leaning on along the last paragraphs: the fictiodesignator 'The-Harry-Lime-of(*The Third Man*)' corresponds to the proper name plus the associated context 'Harry $\text{Lime}_{(\textit{The Third Man})}$'. Even better, 'The-Harry-Lime-of (*The Third Man*, Greene)' corresponds to the name 'Harry $\text{Lime}_{(\textit{The Third Man}, \text{Greene})}$'. Let us generalize: let *W* a fiction work, *A* its author, *N* the name of one of *W*'s characters and *C(W, A)* the convention that governs the use of *N* in *W*. Then

(20) $\text{Ref}(N_{C(W, A)}) \lhd\rhd \text{Ref}((\text{the-}N\text{-de}(W, A))$

In other words, the reference of the proper name N_i in the context *C(W, A)* corresponds ($\lhd\rhd$) to what is referred to by the fictiodesignator 'the-*N*-de(*W, A*)', according to the Goodmanian rendering. This correspondence relation works in both directions. There is a one-to-one mapping that associates each fictiodesignator to a pair formed by a proper name N_i and a context *C(W, A)*.

The correspondence relation captured in (20) allows to shift from a metaphysical perspective, namely from the intentionalist approach, to a semantic one; and to move backwards. (20) clears up the relation between the Intentional Object Theory and the principle *DOS*. The metaphysical theory itself clashes with the principle. However, the intentionalist may reject the charge of revisionism. He can point out that there is a semantic theory, available to his metaphysical stand, which captures the essentials of the Intentional Object Theory. This theory obeys *OAS* and, accordingly, the Grammatical Constraint as well. Hence, metaphysics is not in conflict with semantics. In any case, conflicts could arise within each domain.[31]

2.10 Summing up

In this chapter. I have compared two views of the nature of fiction characters, the Artifact Theory and the Intentional Object Theory. The Inference (§ 1) has supplied the motive to set the comparison's terms, and the comparison is developed in §§ 7–9. The artifactualist's reasons, as articulated in Thomasson's writings, have been put forward in §§ 2–4 and the intentionalist's in §§ 5–7. I have argued that three of the main artifactualist's claims, [T3], [T5] and [T6], are to be, if not rejected, at least

[31] An alternative way of reconciling semantics and metaphysics could be obtained by relativizing the referential use of proper names to communities of bilinguals, as Kenneth Taylor has done. The kind of examples that Taylor bears in mind (see Taylor, 2000, pp. 23ff.) is that of Kripke's puzzle of belief (see Kripke, 1979), which, while being interesting and deserving being developed, substantially differs from mine.

assessed as controversial. On [T3] fiction proper names ('Harry Lime', 'Ofelia', and so on) refer to artifacts, which allow them to fit the principles, *ERE* and *DOS*. According to *ERE*, a proper name *N* refers to an individual *x* if, and only if, *x* exists. *OAS* demands that semantics prevails over ontology, so that decisions concerning the ontology of fiction should depend on how the theory of meaning handles the fiction discourse. The Intentional Object Theory holds intentional objects—and fiction characters are so thought of—only to exist insofar as they are built into imagination acts. Therefore, fiction names are not understood as proper names, but as non-separated parts of functorial expressions. The resulting expressions, which have the form 'the-*N*-of-(*k*)', are designators. Fictiodesignators obey an extended version of *ERE* that applies not only to proper names but to other singular terms as well. It is an open question whether this is enough to accept that fictiodesignators are not in conflict with *ERE*. Fictiodesignators refer to fictional characters, and these are intentional objects. However, do intentional objects exist in a sufficiently substantial sense of 'existence'? This question requires more elucidatory work to be done.

Do intentionalists agree to [T5], i.e. to the claim that fiction names refer to fiction characters? While the artifactualist gives a sharp affirmative answer to this question, the intentionalists do not approve of the idea that fiction names make an onto-semantic contribution of their own, viz. a separate contribution to the truth conditions of the sentences in which those names figure. The contribution will be met by the corresponding fictiodesignators. However, in order for a fiction character to be referred to, it is necessary to involve a novel or a film's plot in the task. Fiction characters' existence thus depends on the work in which their deeds are told: they are intentional objects. In change, Thomasson and artifactualists like van Inwagen and Kripke have somewhat a more relaxed attitude towards their creatures. They believe that artifacts are out there as long as the characters have a foundation in the plots which they inhabit: that is, as long as it be possible to traces current mentions of them by readers and spectators back to a distant written text or a people's narrative tradition. They require in addition that their names either feature in the text or be echoed within the literary tradition.

The Intentional Object Theory contradicts [T6]. First, because artifactualists like Thomasson and Kripke, in abiding themselves by a causal-historical theory of reference, take fiction names to be cathegorematic expressions. This allows them not to come into conflict with *DOS*, thus subjecting their position to the Grammatical Constraint. True, intentionalists do not follow the artifactualists' track in this respect, and this might feed the conclusion that intentionalists disobey *OAS* (and the Grammatical Constraint). Nevertheless, I have argued that semantics and metaphysics do not contradict each other, and that intentionalists' break with *OAS* can be tempered by using the Goodmanian trick to shape a semantic theory of reference that accords with *DOS*. A second reason why intentionalists reject [T6] is that they take the Equivocality Problem seriously. At the same time, they assume that it is impossible to comply with *ERE* and *OAS* and to dodge the Equivocality Problem. In the end, intentionalists accept an extended version of *ERE*, disobey *OAS* and avoid the Equivocality Problem, whereas artifactualists obey *ERE* and *DOS*, but must confront the Equivocality Problem. All in all, the main disagreements between the Intentional Object Theory and the Artifact Theory concern *OAS* and their respective relation to the Equivocality Problem.

References

Anscombe, E. (1965). The intentionality of sensation: A grammatical feature. In R. Butler (Ed.), *Collected papers. Vol. II: Metaphysics and the philosophy of mind*. Basil Blackwell. Originally published in *analytical philosophy: 2nd. Series*. Blackwell.

Bonomi, A. (2008). Fictional contexts. In P. Bouquet, L. Serafini, & R. Thomasson (Eds.), *Perspectives on context*. CSLI Publications.

Crane, T. (2001). Intentional objects. *Ratio, 14*(4), 336–349.

Crane, T. (2010). *The objects of thought*. Oxford University Press.

Crittenden, C. (1991). *Unreality: The metaphysics of fictional objects*. Cornell University Press.

Currie, G. (1990). *The nature of fiction*. Cambridge University Press.

Deutsch, H. (2000). Making Up Stories. In A. Everett & T. Hofweber (Eds.), *Empty names, fiction and the puzzle of non-existence*. CSLI Publications.

Donnellan, K. (1972). Proper names and identifying descriptions. In D. Davidson & G. Harman (Eds.), *Semantics of natural language*. D. Reidel.

Evans, G. (1982). *The varieties of reference*. Clarendon Press.

Evans, G. (1985). The causal theory of names. In *Collected papers*. Clarendon Press.

García-Carpintero, M. (2016). *Relatar lo ocurrido como invención*. Cátedra.

García-Carpintero, M. (2019). Semantics of fictional terms. *Teorema, 38*(2), 73–100.

Geach, P. (1972). *Logic matters*. Blackwell.

Goodman, N. (1976). *Languages of art* (2nd ed.). Hackett Publications.

Goodman, N. (1984). *Of mind and other matters*. (§ III.3.) Harvard University Press.

Howell, R. (2015). Objects of fiction and objects of thought. In S. Stuart Brock & A. Everett (Eds.), *Fictional objects*. Oxford University Press.

Husserl, E. (1913/1952). *Ideen zu einer reinen Phänomenologie un phänomenologischen Philosophie*: Martinus Nijhoff. Translated into English by N. R. Boyce Gibson: *Ideas: General introduction to pure phenomenology*. George Allen & Unwin. 1958.

Ingarden, R. (1964). *Der Streit un die Existenz der welt*. Vol. 1: *Existentialontologie*. Tübingen. Max Niemeyer. Partially translated into English as *Time and modes of being*, Charles C. Thomas Publisher. 1968.

Ingarden, R. (1968). *Vom Erkennen des Literischen Kunstworks*. Max Niemeyer.

Jackendoff, R. (1993). *Semantics and cognition*. The MIT Press.

Johnston, M. (1988). The end of the theory of meaning. *Mind and Language, 3*(1), 28–63.

Kripke, S. (1972). Naming and necessity. In D. Davidson & G. Harman (Eds.), *Semantics of natural language*. D. Reidel.

Kripke, S. (1979). A puzzle about belief. In A. Margalit (Ed.), D. Reidel. Also in *The collected papers of Saul Kripke*. Volume 1: *Philosophical troubles Meaning and use*. Oxford University Press.

Kripke, S. (2013). *Reference and existence*. Oxford University Press.

Lamarque, P. (2003a). Fiction. In J. Levinson (Ed.), *The Oxford handbook of aesthetics*. Oxford University Press.

Lamarque, P. (2003b). How to create a fictional character. In B. Gaut & P. Livingston (Eds.), *The creation of art. New essays in philosophical analysis*. Cambridge University Press.

Lamarque, P., & Olsen, S. H. (1994). *Truth, Fiction and literature*. Clarendon Press.

Lewis, D. (1983). Truth in fiction. In *Philosophical papers. Volume 1*. Oxford University Press. Originally published in *American philosophical quarterly* 15, 1978, 37-46.

Moyal-Sharrock, D. (2009). The fiction of paradox: Really feeling for Anna Karenina. In Y. Gustafson, C. Kronqvist, & M. McEachrane (Eds.), *Emotions and understanding: Wittgensteinian perspectives*. Palgrave Macmillan.

Parsons, T. (1980). *Nonexistent objects*. Yale University Press.

Predelli, S. (2002). «Holmes» and Holmes—A Millian analysis of names of fictions. *Dialectica, 56*(3), 261–279.

Recanati, F. (1993). *Direct reference*. Basil Blackwell.

Recanati, F. (2000). *Oratio Obliqua, Oratio Recta*. The MIT Press.

Reimer, M. (2005). The ellipsis account of fiction-talk. In R. Elugardo & R. Stainton (Eds.), *Ellipsis and nonsentential speech*. Springer.

Russell, B. (1905/1994). On denoting. In *The collected papers of Bertrand Russell. vol. 4: Foundations of logic (1903–1905)*. Routledge.

Saarinen, E. (1979a). Backwards-looking operators in tense logic and in natural language. In E. Saarinen (Ed.), *Game-theoretical semantics*. D. Reidel.

Saarinen, E. (1979b). Intentional identity interpreted: A case study of the relations among quantifiers, pronouns, and propositional attitudes. In E. Saarinen (Ed.), *Game-theoretical semantics* (Vol. 2, pp. 151–223). D. Reidel.

Sainsbury, M. (2010). Intentionality without exotica. In R. Jeshion (Ed.), *New essays on singular thoughts*. Oxford University Press.

Salmon, N. (1998). Nonexistence. *Noûs, 32*(3), 277–319.

Sartre, J. P. (1952). *L'imaginaire: psychologie phénoménologique de l'imagination. Gallimard. Translated into English as The Psychology of Imagination* (p. 1991). Carol Publishing Group.

Schiffer, S. (1994). The paradox of meaning. *Noûs, 28*, 279–324.

Schiffer, S. (1996). Language-oriented, language-independent entities. *Philosophical Topics, 24*(1), 149–167.

Schiffer, S. (2003). *The things we mean*. Oxford University Press.

Searle, J. (1979). The logical status of fictional discourse. In *Expression and meaning*. Cambridge University Press.

Taylor, K. (2000). Emptiness without compromise: A referentialist semantics for empty names. In A. Everett & T. Hofweber (Eds.), *Empty names, fiction and the puzzle of non-existence*. CSLI Publications.

Thomasson, A. (1999). *Fiction and metaphysics*. Cambridge University Press.

Thomasson, A. (2003a). Speaking of fictional characters. *Dialectica, 57*, 2052–2223.

Thomasson, A. (2003b). Fictional characters and literary practices. *British Journal of Aesthetics, 43*(2), 138–157.

Thomasson, A. (2007). *Ordinary objects*. Oxford University Press.

Thomasson, A. (2008). Existence Questions. *Philosophical Studies, 141*, 63–78.

Thomasson, A. (2009). The easy approach to ontology. *Axiomathes, 19*, 1–15.

Thomasson, A. (2013). Fictionalism versus Deflationism. *Mind, 122*, 1023–1051.

Thomasson, A. (2015). Fictional discourse and Fictionalisms. In S. Brock (Ed.), *Fictional objects*. Stuart and Anthony Everett. Oxford University Press.

Thomasson, A. (2018). *Ontology Madre easy*. Oxford University Press.

van Inwagen, P. (1977). Creatures of fiction. *American Philosophical Quarterly, 14*(4), 299–304.

Voltolini, A. (2006). Fiction as a base of interpretation contexts. *Synthese, 153*, 23–47.

Walton, K. (1978). Fearing fictions. *Journal of Philosophy, 75*(1), 5–27.

Walton, K. (1999). *Mimesis As Make-Belief*. Harvard University Press.

Yasigawa, T. (2001). Against creationism in fiction. *Philosophical Perspectives, 15*, 153–172.

Zalta, E. (1983). *Abstract entities*. D. Reidel.

Chapter 3
The Ontology of Fiction: Literature Versus Film

Otávio Bueno

Abstract Amie Thomasson's conception of the ontology of fiction in terms of abstract artifacts provides an illuminating account of the existence and persistence conditions for fictional objects and a sophisticated defense of realism about fictional characters. The conception was primarily articulated with a focus on literature, where perceptual experiences play a relatively small role. How well does the conception fare when we consider artforms that rely crucially on perceptual experiences and perceptual imagination, such as film? In this paper, I examine this issue and raise some concerns about the abstract artifact conception when it is extended to this realm. I then articulate an empiricist alternative that does not depend on realism about fictional characters, and which is able to make sense of both films and literary practices in a principled way.

3.1 Introduction

The ontology of fiction involves, minimally, the specification of the nature of fictional objects: what kind of things they are; their identity and persistence conditions, and whether they exist or not. But no ontological conception of fictional objects, or of any other objects for that matter, can (or should) be entirely divorced from epistemic considerations. Whenever objects are posited as part of a metaphysical conception, whether of fiction or some other area of inquiry, it is reasonable to expect that an account also be provided of how one can come to know—or, at least, form reliable beliefs about—these objects. In the absence of considerations of this sort, concerns can be legitimately raised about the suitability of the ontology in

O. Bueno (✉)
Department of Philosophy, University of Miami, Coral Gables, FL, USA

Graduate School of Information Sciences, Tohoku University, Sendai, Japan

M. Sambrotta (ed.), *Metaphysics Today: In Conversation with Amie Thomasson*, Synthese Library 512, https://doi.org/10.1007/978-3-031-98468-6_3

question. Certain characterizations of the domain of fictional entities may turn into a mystery how one can come to know these objects; alternatively, suitable ways of knowing them may be insufficient to account for how true claims about fiction can be properly made.

This is an aspect of what Christopher Peacocke raised as the integration challenge: the problem of properly reconciling one's ontology with one's epistemology. As he notes:

> The problem of reconciliation may take various forms. We may have a clear conception of the means by which we ordinarily come to know the statements in question. Yet at the same time we may be unable to provide any plausible account of truth conditions knowledge of whose fulfilment could be obtained by these means. Alternatively we may have a clear conception of what is involved in the statements' truth, but be unable to see how our actual methods of forming beliefs about their subject matter can be sufficient for knowing their truth. In some cases we may be unclear on both counts. I call the general task of providing, for a given area, a simultaneously acceptable metaphysics and epistemology, and showing them to be so, the *Integration Challenge* for that area (Peacocke, 1999, p. 1).

Amie Thomasson developed an elegant approach to the ontology of fiction in terms of abstract artifacts (Thomasson, 1999). Along the way, her account also offers identity and persistence conditions for fictional objects, a positive answer to the question regarding their existence, and a principled strategy to approach the resulting integration challenge. After all, if fictional objects do exist, an account needs to be articulated of how one can acquire knowledge of them, and according to Thomasson, provided that they are understood as abstract artifacts, fictional characters pose no difficult at all for how we can form reliable beliefs about and even have knowledge of them.

In what follows, I assess how well abstract artifacts can accommodate knowledge of or, at least, justified belief about fictional characters in the context of artforms for which a particular kind of perceptual experiences—perceptual imagination (Scruton, 1979)—play a crucial role. Perceptual imagination consists in focusing and restricting one's imaginative acts to what is being perceived in a given context. Films provide significant instances of such artworks, since watching a film involves perceiving the content of what is projected on a screen and suitably adjusting one's imaginative acts to that content. The experience of watching a film, on a realist account, is one of seeing fictions: one has access to the world of the film by seeing the fictional characters in the various scenes that compose the motion picture. The challenge is then how the abstract artifact theory can explain the phenomenology of film—what it is like to watch a film—since one cannot see an abstract object. After all, abstract objects are not located in space (or, more generally, in time). If fictional characters are not seen, it is unclear how the phenomenology of film can be accounted for by abstract artifact theorists. Motivated by this difficulty, I advance an empiricist alternative that does not rely on the existence of fictional characters, and which allows one to accommodate both films and literary practices in a principled way.

3.2 Abstract Artifacts: Literature

The distinctive trait of Thomasson's (1999) abstract artifact approach to fiction is the way in which it allows for the combination of realism about fictional characters—the claim that such characters do exist—with a sensible epistemology of how we can have knowledge of or, at least, form reliable beliefs about them. On this view, fictional characters are created by the intentional acts of those who brought them about. All that it takes for Tolstoy to create Anna Karenina is for him to describe her and her deeds, by indicating her properties and her actions. Readers have access to Anna Karenina by reading what is explicitly stated in the novel and by determining what follows from what is so stated in the work.

Abstract artifacts are artifacts since they are created by the intentional acts of their authors and are abstract since they lack spatial location (they do have temporal properties, since they are created at a particular moment in time and may no longer exist at another). Those wondering in Russia in the final quarter of the nineteenth century would have never met Anna Karenina and would be confused even to search for her in the streets of Russian cities.

According to Thomasson (1999), abstract artifacts have both existence and persistence conditions. Existence conditions are those that determine what it takes for fictional characters to come into being, that is, to exist, whereas persistence conditions specify what it takes for such characters to remain in existence. As noted, abstract artifacts have very minimal existence conditions: the intentional acts of an author, describing the properties and the deeds of a fictional character, for instance, in the context of writing a novel or a short story in which it features—are all it takes for such character to come into existence. The character will remain in existence if access to the relevant descriptions is secured by the reading of copies of the books in which the fictional character is described or by having memories of the corresponding stories. Provided that these concrete objects are in place—books, memories, and similar resources—the correlated fictional characters will remain in existence. Otherwise, they will no longer exist.

Given their existence and persistence conditions, fictional characters are temporal objects: before their creation, they do not exist; as soon as their persistence conditions fail, they no longer do. They may come back into existence if the concrete objects on which they depend are retrieved. However, if someone else reinvents the stories, a different fictional character is involved. In this respect, the view requires a significant form of intentionalism, in which the identity of a character depends on the intentional acts of its author (or authors). As Thomasson notes:

> The immediate dependencies of a fictional character are first, on the creative acts of its author or authors, and second, on a literary work. Clearly the dependence of a fictional character on the intentional acts of its creator or creators is a rigid historical dependence (Thomasson, 1999, p. 35).

Considering these features, the abstract artifact theorist is able to explain how beliefs about, and knowledge of, fictional characters can be reliably formed in a perfectly straightforward and clear way. By reading *Anna Karenina*, we find out about Anna's features and her actions, her motives, and her fears. Whatever is explicitly stated in the work about her, or follows logically from what is so stated, are features of her and in this way, readers form beliefs about her and acquire knowledge of her properties. The work may not settle several issues about her and about other characters in the novel. These items are just left undecided.

Crucial to the abstract artifact approach is the role of imagination in both the creation and in the reception and the corresponding dissemination of fiction. Intentional acts of creation of a fictional character are imaginative acts: authors imagine their characters, animating them with their features and describing their deeds. Similarly, the reception and dissemination of fiction also involves imagination as readers imagine the characters that are described in the relevant novels and short stories or that are conveyed in stories that are told over time.

The resulting approach is elegant in its minimalism about fictional characters and in the way that it demystifies the epistemology of fiction within a realist view about fictional characters. One can thereby posit the existence of fictional characters with the assurance of having an account of how one can have knowledge of, and form justified beliefs about, them. In this way, the integration challenge—of providing a suitable epistemology given a particular ontology of fictional characters—can be met.

The abstract artifact approach works especially well in the context of literature, since only imagination is needed to provide access to the relevant characters. Perceptual experiences play virtually no role in the account. The issue then arises as to how well the approach fares when the focus shifts to artworks for which perceptual experiences are decisive to the characterization and dissemination of the relevant works. This is the case of films. In what follows, I will examine them.

3.3 Transparency in Film

Films are significant in the way they bring together perceptual—visual and auditory—experiences of recorded action as the crucial means of access to the resulting motion pictures. The fact that, to be created, films involve a plethora of activities—including script writing, acting, directing, sound engineering, and editing, among others—clearly highlights their compositional complexity. The experience of films is similarly intricate, with multisensory exposure to visual and auditory content, together with associated imaginative acts, being required for the apprehension of the work.

Since the content of films results from the recording of certain acts, the corresponding moving images become crucial. This raises the issue regarding the connection between the images that, in part, compose a film (a motion picture) and the objects and events whose images are displayed in a film. There are those who claim that this relation is one of transparency: in seeing the images, one sees the objects and events that are depicted in them. The images are, in this respect, transparent (Walton, 1984, 2008).

Nonetheless, are motion pictures transparent? What does transparency ultimate amount to in the context of films? If movies are transparent, what is the significance of this feature to the ontology and the phenomenology of film? As Gregory Currie notes—and in the discussion below I follow much of his views—to the extent that films rely on photography, reproducing rather than just representing objects in the actual world, the resulting images are transparent. On his view:

> Following Kendal Walton (1984), I shall call this the doctrine of *Transparency*: film is transparent in that we see 'through' it to the real world, as we see through a window or a lens (Currie, 1995, p. 20).

The issue then arises: what do we see when we watch a film?

Before examining this issue, it is important to distinguish *transparency* from a related, but importantly different, notion of *counterfactual dependence* in perceptual experience (which, ultimately, characterizes a central trait of vision):

(C1) Had the scene before one's eyes been different (within the sensitivity range of the eyes), the resulting perceptual experience would have been correspondingly different (see Lewis, 1980).

(C2) Had the scene before one's eyes been the same (within the sensitivity range of the eyes), the resulting perceptual experience would have been correspondingly the same (for a discussion, see Bueno, 2018b).

The counterfactual dependence between the scene before one's eyes and the resulting perceptual experiences allows one to track the environment around us by using one's own perceptual experience. It allows us to have access to objects that is *robust* (what is seen does not depend on what is believed), can be *refined* (one can get closer for a better look), and allows us to *track* the relevant objects in space and time (Azzouni, 2004; Bueno, 2005, 2018b).

In movies, counterfactual dependence can be formulated as follows:

(C1*) Had the scene before the camera been different (within the sensitivity range of the camera), the resulting image would have been correspondingly different.

(C2*) Had the scene before the camera been the same (within the sensitivity range of the camera), the resulting image would have been correspondingly the same.

Due to such a counterfactual dependence, the camera becomes a recording device that is sensitive to the visually salient features of the objects it records. Of course, these conditions hold during the recording of a film. Editing and computer-generated images, among other interventions on a film, may violate them.

Sometimes counterfactual dependence and transparency are perilously mixed. Although Noël Carroll (2008) does not endorse the transparency of movies, at one point he defines transparency in terms of counterfactual dependence:

> So, x is a transparent image if and only if (1) x puts us in mechanical contact with its object in such a way that the visible output in x is counter-factually dependent upon its visible input, and (2) x preserves genuine or real similarity relations between the objects and events presented by it. Furthermore, cinematic images are transparent in this sense. Cinematic images are like telescopes into the past. We see through them to the sound stages of yesterday (Carroll, 2008, p. 97).

One can see why it is tempting to formulate transparency in this way. If the counterfactual dependence between the scene before the camera and the resulting image obtains, one may be tempted to say that in seeing the image we thereby see "through" it to the scene that generated it.

However, this is not quite right. In seeing the image of an object on the screen, we literally see just an image, not the object the image represents. The object may not exist anymore, and so it could not be seen at that point. And even if it did exist, no longer would it be accessible when its image is finally seen. When I see an image of Harrison Ford in *Blade Runner*, I do not see the actor, since he is not an image, but a man. I see a *recording* of Harrison Ford, which clearly is *different* from Harrison Ford. Counterfactual dependence is an important feature of perceptual experience and recording instruments, such as cameras or microscopes. But it falls short of securing transparency. One may *infer*, given the way in which the image was produced, that it was generated by objects having the visually salient features displayed on the image. However, this is different from seeing the objects, since the image of an object is different from the object itself—for instance, the former is two-dimensional, the latter is three-dimensional. When an image is seen, what is seen is not its object, but just the corresponding image (for further discussion, see Currie, 1995). Of course, there is an important visual similarity between the object and its image—a pic of grandma does look like grandma, at least in significant respects. As a result, considering the way in which the image was produced, image and object are often conflated. Not surprisingly, a photograph of grandma is often presented as being grandma, as when it is said: "Look [pointing to the photograph], that's grandma!" Obviously, one's grandmother is not a photograph.

Given these considerations, when we watch a film, what we ultimately see are images: the images of the actors when the film was shot, disregarding computer-generated images (Currie, 1995). Yet, in the case of fictional films, we should see *characters*, parts of the fiction that is being displayed. Nevertheless, how can we possibly *see* fictions in film if fictional characters do not exist and we can only see things that do exist—in spacetime? Different answers are offered depending on the ontological conception under consideration.

On a view that acknowledges that fictional characters do exist—an ontologically committing view such as Thomasson's (1999)—in the case of film, presumably all it takes for one to see a fictional character is for one to watch the screening of a film. This would be equivalent of telling a story about a character, but instead of doing it in writing, the story is told with images. One would then see the characters on film by seeing the images on the screen. (See the discussion of the somewhat ambiguous notion of *imagined seeing* in Wilson, 2011.) However, on the artifactual view, fictional characters are abstract objects (abstract artifacts)—that is, non-spatially located, but temporal, entities, which are created in time, and may fail to remain in existence also in a particular moment in time. This raises the issue: given fictional characters' abstract nature, how can one possibly *see* them? Only objects that are in space and that reflect light can be seen. As a result, it then becomes mysterious how is it possible to see fictional objects. Given that seeing fictions in film is arguably a significant aspect of one's experience of (fiction) motion pictures, an account that seems unable to accommodate this trait is significantly incomplete.

In contrast, on an ontologically non-committing view, fictional objects do not exist. Clearly, we cannot see these objects, since existence in spacetime is a necessary condition for an object to be seen. Nevertheless, what is gained in ontological parsimony—by not requiring the existence of fictional objects—is lost in phenomenological inaptness. For it becomes unclear how the resulting view can make sense of the phenomenology of watching a film, since when we watch a motion picture, it seems that we see characters in film. However, given that we are unable to see what does not exist, and since on this view fictional characters do not exist, we cannot see them. As a result, making sense of the phenomenology of experiencing a film does not seem to be possible also on an ontologically non-committing view (see also Currie, 1995; Bueno, 2018a).

3.4 Abstract Artifacts in Literature and Film: Ontology, Phenomenology, Experience

It was noted that the abstract artifact approach is especially apt for literature. There are, however, significant differences between literature and film. Literature relies primarily on the readers' imagination, their capacity to engage with the content of a literary work and conjure up the various scenarios and situations that are described or evoked by that work. Thus, linguistic representation is the primary device readers employ to access the work's content. Perceptual experiences play a very limited role. Those who engage with literary works of course read texts or listen to them but, typically, apart from literary forms such as concrete poetry, in which pictorial elements do have significance, perceptual experiences are not at all central to provide distinctive access to the content of the work. They play no crucial function. In contrast, imaginative acts do much of the heavy lifting of making the content manifest to the reader. No matter how detailed a scene is described in a novel, there will always be significant elements that are left to the reader to fill in with imaginative acts.

In *The Book of Disquiet*, an intriguing literary reflection about the self and personal identity, Fernando Pessoa writes:

> I was a foreigner in their midst, but no one realized it. I lived among them as a spy and no one, not even I, suspected it. They all took me for a relative; no one knew I'd been swapped at birth. And so I was one of their equals without anything in common, a brother to all without belonging to the family (Pessoa, 2015, text 433; see also Zenith, 2021).

What is the relationship between this self that Pessoa describes and the community he may (or may not) belong to? He describes himself as "one of their equals", a "spy", a "brother", but immediately takes back the descriptions. No brother can fail to "[belong] to the family" (a brother is always a brother no matter what the family may say on the matter); no spy can be a spy without suspecting that one is a spy; no one can be "one of their equals" without having "anything in common" with them. Which self is, in the end, described in this passage? This self is a "foreigner" not only to the community in question but primarily to himself.

The reader can only wonder what is going on, imagining scenarios about the possible identity of this self, ultimately unable to determine that identity, which perhaps is not transparent even to the author. Characters whose identity is deliberately indeterminate may raise a challenge to the abstract artifact view. Being significantly underspecified, the description of the character fails to delineate which person is being characterized in the relevant passage. It becomes unclear which fictional character, after all, one is supposed to grasp. It is not clear which character, in fact, is posited by the author. This requires an even larger role for imagination, constantly toying with very different possible selves who would be picked out by the deliberately understated descriptions.

In other cases, the descriptions are fully detailed, but still much is left open for imagination. In Paul Auster's *The Book of Illusions*, vivid descriptions are offered of silent films that Hector Mann created in the beginning of the twentieth century. Here we have a (fictional) character making (nonexistent) films, which are then glowingly described. Hector not only directed these silent films, but he was also the leading actor in them, captivating the audience with his ingenious acts. As Auster notes:

> Before the body, there is the face, and before the face there is the thin black line between Hector's nose and upper lip. A twitching filament of anxieties, a metaphysical jump rope, a dancing thread of discombobulation, the mustache is a seismograph of Hector's inner states, and not only does it make you laugh, it tells you what Hector is thinking, actually allows you into the machinery of his thoughts. Other elements are involved—the eyes, the mouth, the finely calibrated lurches and stumbles—but the mustache is the instrument of communication, and even though it speaks a language without words, its wriggles and flutters are as clear and comprehensible as a message tapped out in Morse code.
>
> None of this would be possible without the intervention of the camera. The intimacy of the talking mustache is a creation of the lens. At various moments in each of Hector's films, the angle suddenly changes, and a wide or medium shot is replaced by a close-up. Hector's face fills the screen, and with all references to the environment eliminated, the mustache becomes the center of the world. It begins to move, and because Hector's skill is such that he can control the muscles in the rest of his face, the mustache appears to be moving on its own, like a small animal with an independent consciousness and will. The mouth curls a bit at the corners, the nostrils flare ever so slightly, but as the mustache goes through its antic gyrations, the face is essentially still, and in that stillness one sees oneself as if in a mirror, for it is during those moments that Hector is most fully and convincingly human, a reflection of what we all are when we're alone inside ourselves. These close-up sequences are reserved for the critical passages of a story, the junctures of greatest tension or surprise, and they never last longer than four or five seconds. When they occur, everything else stops. The mustache launches into its soliloquy, and for those few precious moments, action gives way to thought. We can read the content of Hector's mind as though it were spelled out in letters across the screen, and before those letters vanish, they are no less visible than a building, a piano, or a pie in the face (Auster, 2002, pp. 29–30).

One can almost see the scenes that Hector has concocted and how his mustache vividly displays the content of Hector's mind. In Hector's able hands (as elegantly described by Auster), silent movies do not need words to be as transparent as explicit close captions; they require no words to have their content made as visible as a pie in the face. An interesting layer of descriptions is involved here as a fictional character creates films that no one has ever seen, and from the vibrant descriptions of these films, readers have a glimpse of how they looked like without ever seeing a single of their frames.

No matter how the scenes are arrestingly described by Auster, readers still need to provide a lot of content to fill in the details. The texture of previously watched silent films from the 1920s are likely to be imported to imagine the way Hector's films may have looked like; the exact proportion of the screen that his mustache occupies in a close shot also needs to be specified and, hence, imagined, and so does the mustache's approximate thickness. All these items would have been immediately visible if a single frame of Hector's mustache could be displayed. But one cannot show, nor can one see, what does not exist. The result is crucially to rely on the imagination to complete, as much as possible, what cannot be fully specified with descriptions alone.

In contrast, films engage directly and primarily with the visual and auditory modalities of viewers. Nothing similar in relevance or scope regarding perceptual experiences is found in literature. The primary mode through which content is conveyed to film viewers is *via* visual and auditory representations. Central to one's access to a film's content is a proper phenomenology: the experience of viewing the film. This involves seeing the images that are projected on the screen, listening to the accompanying sounds, and integrating the resulting experiences into a coherent whole. Phenomenologically, this amounts to constituting the film as the work is experienced by the viewer (Casebier, 2009).

There is also indeterminacy in film. David Lynch's *Mulholland Dr.* provides a clear illustration. At least two substantially different interpretations of the film are sustained by the work. (i) One interpretation gives priority to the dream content of the work. On this view, the first part of *Mulholland Dr.*—which lasts from the beginning of the film until the end of the scene in which the Theater Silencio appears—is composed in its entirety by a dream that the actor Diane Selwyn (which is played by Naomi Watts) has (see Vaida & Wildman, 2005). In contrast, in the second part of the film, the brutal reality of the world becomes manifest to Diane, who must face the loss of her lover, Camilla Rhodes (played by Laura Harring) and confront the failure of her own career.

(ii) A different interpretation reverses the direction of fit. On this view, priority is given to reality and the ordering of what is real or dreamt is flipped. In the film's first part, the events portray Betty's life, and her aspiration to become an actor in Los Angeles, having just arrived in the city for an audition. In contrast, the film's second part—beginning with the scenes involving the Theater Silencio—portrays one of Betty's worse nightmares, namely, to fail as an actor and to lose her lover, the successful actor Camilla Rhodes. Unable to cope with this scenario, Betty decides to kill her former lover.

Mulholland Dr. is constructed in such a way that nothing in the work settles the issue of which of these widely divergent interpretations is correct. Both interpretations are, in fact, sustained by the work—among others that are also possible. As a result, the film is genuinely and intentionally designed to be an indeterminate object, which allows for a multiplicity of interpretations. It is a clear instance of Umberto Eco's open work (Eco, 1989). Particularly significant is the multi-level indeterminacy that characterizes *Mulholland Dr.*—at the levels of character, narrative, and interpretation. It is an inherently indeterminate work.

While the role of imagination in literature is largely unproblematic for the abstract artifactual view, given the constraints on imagination imposed by the perceptual experiences of a film, it becomes less clear how the view is supposed to accommodate the phenomenology of motion pictures. In particular, it is unclear how the indeterminacy of film can be accounted for, since the indeterminacy is displayed *via* the visually salient content of the work, the images that allow for one's access to the film's content. Furthermore, fictional characters are not images on a screen—they are supposed to be agents doing their deeds in the world of the film—and, as noted above, it is also unclear how such characters could possibly be seen if they are abstract entities, which are not spatially located anywhere. An alternative perspective seems to be called for.

3.5 An Empiricist Alternative

Central to the proposal suggested here is a form of empiricism, including a direct realist approach to perceptual experiences (Bueno, 2018b, 2024). Empiricism is here understood as a stance, a form of investigating the world rather than a cluster of doctrines about the nature of experience and reality (see van Fraassen, 2002). This allows the empiricist to remain largely agnostic about several issues that more substantive metaphysical views aim to settle—including the existence of fictional characters and abstract objects more generally. Settling these issues is ultimately not required to make sense of film or literature as our literary and cinematic practices are largely undisturbed by the way in which these metaphysical issues are resolved (or not).

For instance, understanding what is real or imagined in a film such as Alfred Hitchcock's *Vertigo* is a key trait in making sense of the film. But no matter how one understands the fictional characters involved—as existing abstract artifacts or as nonexistent results of the viewers' imaginative acts—it does not change the puzzling elements that the film aptly articulates. The same goes for *Mulholland Dr.* Regardless of the way in which Betty, Diane, and Camilla are ultimately characterized—whether in terms of abstract objects or concrete imaginings—understanding the film demands making sense of which aspects of the narrative should be taken as actually taking place in the world of the film and which are only dreamt.

Crucial to this understanding is the fact that these characters are apprehended *via* perceptual experiences. It is by seeing the images that constitute the film that one has access to the film's content. As noted above, the film is about the characters involved—even if they happen not to exist. To accommodate the experience of watching a film (the phenomenology of experiencing the work) one needs to realize that a special form of imagination is involved: perceptual imagination. This is an imaginative act constrained by the perceptual content of a given perceptual experience. While watching *Blade Runner* and seeing images of Harrison Ford, one is supposed to image that these images are not of Ford but rather of Rick Deckard, whose job is to track down replicants and "retire" them. What one imagines is constrained by the content of what is displayed on the screen. One is supposed to imagine Deckard as looking and behaving like Ford does. The imaginative act is constrained by the content of what is experienced by watching the screening of the motion picture. (For further discussion, see Scruton's (1979) study of imaginative perception, and especially the view defended in Currie (1995); see also Bueno (2018a).)

Direct realism, understood in a non-dogmatic way (Bueno, 2024), indicates that the immediate objects of visual perception are objects and their properties that fall under one's visual field. Applied to film, the account indicates that one sees images projected on the screen. An imaginative act then takes these images as standing for the fictional characters in the world of the film. Once this act is implemented, viewers follow along the narrative portrayed on the screen, immersing themselves in the deeds of the characters. In this way, even though fictional characters cannot literally be seen, since they do not exist (or need not be taken to exist), the phenomenology of film can be accommodated in terms of perceptual imagination, the combination of an imaginative act constrained by the perceptual content that is experienced watching the film (see Currie, 1995).

It may be argued that the abstract artifact approach could easily adopt the same move to make sense of the phenomenology of film, since the view already recognizes the role of imagination in dealing with fictional characters. The concern, however, is that these characters are taken to exist and although readers do not have to see such characters in literature, in the case of film the perceptual content is crucial for any account of the phenomenology of film experience. As a result, it becomes difficult to harmonize the commitment to the existence of (unobservable) abstract artifacts (since they are not spatial entities) with one's perceptual access to the film's content. In the end, in light of the significant difference between the experience of literature and film—given the role of imagination in the former and perceptual imagination in the latter—a view that grants proper prominence to the perceptual content of films fares better than one that does not.

With perceptual imagination in film, it is possible to accommodate one's dealing with fictional characters, even though they cannot be seen, given their nonexistence (or, at least, given that their existence is not required to make sense of motion pictures). In contrast with Currie (1995), however, the engagement with fictional characters need not be a form of simulation. After all, viewers often respond directly and, where appropriate, *empathically* toward the characters in a film rather in terms of indirect simulations. The requirement of simulation seems to add an additional layer of processing that removes viewers from their immediate reactions to the film's content.

This immediate (unmediated) reaction to a film is highlighted both by the direct realism involved in watching the work and by the realistic traits of the medium. In a certain respect, every motion picture is realistic—with the obvious exception of computer-generated ones. After all, a motion picture is the result of whatever was recorded when the movie was being shot, and in this respect, the counterfactual dependence between the scene before the camera and the resulting image plays a significant role in the construction of a film. Despite editing and the addition of music, what is seen on the screen is the result of what was in front of the camera when the relevant scenes were recorded (for discussion, see Carroll, 2008).

In another respect, nevertheless, motion pictures are *not* realistic. They are the result of edits, contrastive comparisons by the juxtaposition of certain scenes, and suggested connections that often defy expectations of how events actually unfold. Consider, for instance, the sequence toward the end of Paul Thomas Anderson's *Magnolia*, a typical realistic film—despite the biblical rain of frogs—in which the various characters, trapped inside their personal and family drama, suddenly start singing Aimee Mann's song "Wise Up". A crucial role of perceptual imagination is found in this context, the shared song brings together the plight of the various characters, and suddenly, despite their inadequacies and the many challenges that each of them face, they are all living together their lives and responding, however they can, to whatever fate has thrown at them (frogs and all!).

This approach is also able to accommodate the indeterminacy of film. The indeterminate content of a work, such as *Mulholland Dr.*, is the result of the deliberate editing of the film that prevents the framing of the scenes to determine what is real and what is only being dreamt in the world of the film. For instance, no indication (no mark) is made in the work of what has actually happened and what has not. Thus, it is indeterminate how to respond to each scene and the indeterminacy of the work emerges. The result of the editing is that seeing directly the scenes portrayed on the screen is not enough to determine how the film's content should be interpreted. The outcome is an inherent indeterminacy.

Finally, the issue of seeing fictions in film can similarly be addressed. After all, an imaginative act of constraining the content of the imagination by what is displayed on the screen allows us to experience what the fictional characters in the world of the film are going through. This is accomplished by imagining that the deeds displayed by the images of actors and happenings on the screen are, in fact, actions of characters and the unfolding of events that take place in the world of the film. No commitment to the existence of characters is required to make this point, for the content is just (perceptually) imagined.

3.6 Conclusion

Significant differences between film and literature have been highlighted throughout this essay: in particular, the central role that perceptual imagination plays in the former but not in the latter has been emphasized. Some concerns were then raised to the extension of the abstract artifacts approach to film, given the significance of the perceptual content to the phenomenology of film and the tension that this has with the existence of abstract, and hence unobservable, artifacts. In the end, an empiricist alternative was sketched that allows one to make sense, at least in principle, of our experience of films without having to be committed to the existence of fictional characters.

Acknowledgments For many helpful discussions on the issues examined in this work, my thanks go to Enrique Aranda, Diane Carson, Allan Casebier, Greg Currie, Linda Ehrlich, Steven French, María José García Encinas, and Amie Thomasson.

References

Auster, P. (2002). *The book of illusions*. Henry Holt.

Azzouni, J. (2004). *Deflating existential consequence: A case for nominalism*. Oxford University Press.

Bueno, O. (2005). Dirac and the dispensability of mathematics. *Studies in History and Philosophy of Modern Physics, 36*, 465–490.

Bueno, O. (2018a). Interpreting the sciences, interpreting the arts. In O. Bueno, G. Darby, S. French, & D. Rickles (Eds.), *Thinking about science, reflecting on art: Bringing aesthetics and philosophy of science together*. Routledge.

Bueno, O. (2018b). Empiricism. In J. Saatsi (Ed.), *Routledge handbook on scientific realism* (pp. 96–107). Routledge.

Bueno, O. (2024). How to be a direct realist. In R. French & B. Brogaard (Eds.), *The roles of representation in visual perception* (pp. 251–261). Springer.

Carroll, N. (2008). *Philosophy of motion pictures*. Blackwell Publishing.

Casebier, A. (2009). *Film and phenomenology: Towards a realist theory of cinematic representation*. Cambridge University Press.

Currie, G. (1995). *Image and mind: Film, philosophy and cognitive science*. Cambridge University Press.

Eco, U. (1989). *The open work*. (Translated by Anna Cancogni.). Harvard University Press.

Lewis, D. (1980). Veridical hallucination and prosthetic vision. *Australasian Journal of Philosophy, 58*, 239–249.

Peacocke, C. (1999). *Being known*. Clarendon Press.

Pessoa, F. (2015). *The book of disquiet*. (Translated and edited by Richard Zenith). Penguin.

Scruton, R. (1979). *The aesthetics of architecture*. Princeton University Press.

Thomasson, A. (1999). *Fiction and metaphysics*. Cambridge University Press.

Vaida, I. C., & Wildman, V. H. (2005). Mulholland Dr. *Psychoanalytic Psychology, 22*, 113–119.

van Fraassen, B. C. (2002). *The empirical stance*. Yale University Press.

Walton, K. (1984). Transparent pictures: On the nature of photographic realism. *Critical Inquiry, 1*, 246–277.

Walton, K. (2008). *Marvelous images: On values and the arts*. Oxford University Press.

Wilson, G. (2011). *Seeing fictions in film: The epistemology of movies*. Oxford University Press.

Zenith, R. (2021). *Pessoa: A biography*. Norton and Company.

Chapter 4
Triviality and Solipsism: Two Problems in the Metaontology of Art and a Goodmanian Solution

Nemesio García-Carril Puy

Abstract Amie Thomasson's descriptivism and Julian Dodd's revisionism are two outstanding opposed views within the methodological debate in the metaontology of art. In this paper, I argue that none of them provide an adequate methodology for the ontology of artworks insofar as they do not secure ontological accounts to play a substantive role in improving our understanding of artworks and the practices they involve. Thomasson's descriptivism faces a problem of triviality, unjustifiably reducing art ontology to a trivial task. In turn, Dodd's revisionary approach faces a problem of solipsism, being a methodology that does not ensure conceptual mediation between the ontology of art and artistic practices. Finally, I will sketch a solution for both problems based on Nelson Goodman's distinction between entrenchment and familiarity of a predicate.

Keywords Metaontology · Ontology of art · Descriptivism · Revisionism · Substantivity

4.1 Introduction

The methodological debate in the metaontology of art tries to answer the question of what is the methodology to be followed when doing art ontology. Two main views can be identified within this debate. On the one hand, metaontological descriptivism defends the idea that the task of art ontology is to describe in a systematic way the pre- theoretical ontological conception shared by the agents engaged in our artistic practices. Amie Thomasson supplies one of the most salient

N. G.-C. Puy (✉)
Department of Philosophy and Society, Complutense University of Madrid, Madrid, Spain
e-mail: nemesiog@ucm.es

M. Sambrotta (ed.), *Metaphysics Today: In Conversation with Amie Thomasson*, Synthese Library 512, https://doi.org/10.1007/978-3-031-98468-6_4

accounts within this approach. On the other hand, metaontological revisionism is the thesis that artistic practices and their agents' intuitions can be revised in light of the best ontological proposal about the artworks concerned in terms of theoretical virtues, e.g. simplicity, explanatory power and coherence with our knowledge in other domains. Julian Dodd offers one of the most sophisticated accounts within this approach.

In this paper, I argue that Thomasson's and Dodd's views do not offer adequate methodologies for the ontology of art. It will be shown that their methodologies do not guarantee that ontological accounts may play a substantive role in the enterprise of understanding art objects and artistic practices. Both methodologies may validate ontological accounts with no epistemic value for understanding artworks and the artistic practices they involve. Their methodologies are faced with two problems, respectively, that prevent them from granting the achievement of this goal. Thomasson's descriptivism faces a *problem of triviality*, unjustifiably reducing art ontology to a trivial task. Dodd's revisionism faces a *problem of solipsism*, being a methodology that does not guarantee a conceptual mediation between the outcomes of ontological inquiry and the ordinary view of artworks embedded in our artistic practices.

I will justify in the next section why displaying understanding about artworks and the practices they involve is a desideratum for an adequate methodology for art ontology. In the third section, the main points of Thomasson's descriptivism will be introduced. In the fourth section, we shall see how her view is affected by a problem of triviality. Dodd's revisionism will be introduced in the fifth section, exploring the extent to which it falls under a problem of solipsism. Finally, I will present a methodological solution for both problems based on Goodman's distinction between entrenchment and familiarity of a predicate.

4.2 The Epistemic Value of the Ontology of Art

Reflecting about the task of philosophy, Wilfrid Sellars made this famous metaphilosophical claim: 'THE aim of philosophy, abstractly formulated, is to understand how things in the broadest possible sense of the term hang together in the broadest possible sense of the term' (Sellars, 1963/1991, 1). According to Sellars, the main goal of philosophy is to display *understanding* about the interrelated things of our environment, i.e. about the world. This understanding results in *knowing how* to orient ourselves in our environment from a whole picture of reality. Apart from other existential, rationalistic or normative explanations (the latter is the privileged by Sellars), the need of this sort of understanding is also justified in purely naturalistic terms. Natural evolution has equipped human beings with a linguistic and conceptual apparatus to survive in the absence of capacities that other non-linguistic animals have. To this extent, the orientation provided by this sort of understanding is basic to make predictions that are crucial for the survival of our species.

The need of understanding the world justifies and explains why human beings do philosophy. The holistic understanding of reality is present in all forms of human knowledge. It is present at the level of the everyday knowledge to orientate us in our everyday practices (Sellars, 1963/1991, 3). It is also present at the scientific level to orientate the scientist within her special discipline (Sellars, 1963/1991, 2). However, the scientist, as well as any agent of any specialized practice, demands a sense of how her topic, methods and principles are articulated in 'the countryside as a whole'. This requires a reflective holistic insight that is usually hampered by the specialized knowledge of a scientific discipline or specialized practice, and by the dynamics of our everyday practices. This is precisely the insight that must be supplied by the philosopher. It constitutes the reason of why this understanding, which goes beyond the understanding found at the scientific level or at the level of our everyday and specialized practices, is the ultimate goal of philosophy. Consequently, the epistemic value of philosophy relatively to understanding the world along the lines sketched above goes beyond truth or carving the world at its joints, in contrast to the epistemic value of particular scientific disciplines. It involves joint carving and truth about the world, but does not reduce to them. This notion of 'epistemic value' is broader than the defended one by Theodore Sider (2011, 61), since it is *not* 'better to think and speak' *only* 'in joint- carving terms' for the purpose of a holistic understanding of the world.

If this kind of understanding is the main aim of philosophy, and if ontology is a philosophical discipline, ontology must be also oriented towards this goal. This is a point that recent neo-Sellarsian views have finely highlighted (cf. Cumpa, 2014, 2018, 2020; Buonomo, 2017; Puy 2021). Javier Cumpa acknowledges that human beings have an inescapable 'disposition to understand what is going on in our different environments' (Cumpa, 2020). According to this disposition, 'our understanding as persons, our understanding as scientists, and our understanding as metaphysicians forms a one-to- many relation between our understanding and our practical, scientific, and metaphysical aims' (Cumpa, 2020). To this extent, whichever is the fundamental question assigned to ontology, that question is only substantive relatively to the fundamental goal of understanding the manifest and the scientific images of the world and the relationships between them. For instance, as Cumpa notes, if from a neo-Carnapian approach we consider that the fundamental question of ontology is about the existence of things within a certain framework, this question is substantive only if it has epistemic value for understanding either the manifest image or the scientific images or their relationships. Alternatively, if from a neo-Aristotelian approach we consider that the fundamental question of ontology is about the categorial structure of the world, this question is substantive only if has epistemic value for understanding either the manifest image or the scientific images or the relationships among them.

I will assume that the neo-Sellarsian point of view concerning general ontology has also a fruitful application to the particular domain of the ontology of art. Guided by this aim of understanding in the sense just introduced, a metaontological criterion of substantivity for the ontology of art follows. The questions of the ontology of art are substantive only if they epistemically contribute to our understanding of artworks and the artistic world. In other words, they are substantive only if they have epistemic value for knowing how to orient ourselves to properly appreciate artworks of different kinds and interact with them. The question about the existence of artworks of a certain kind, or the question about whether they are types, sets or fusions, are substantive only if they help us to better understand the artworks of that kind and the practices to produce and properly appreciate them. The epistemic value of art ontology is the goal that justifies this subdiscipline of ontology as a substantive enterprise. Consequently, an adequate methodology must secure that art ontological accounts satisfy this desideratum. Crucially, as we shall see, Thomasson's descriptivism and Dodd's revisionism are methodologies that do not ensure the achievement of this goal. Thomasson's descriptivism and Dodd's revisionism face two different problems, triviality and solipsism, respectively, that lead them to fail this condition. They are to be considered, to this extent, as non-suitable methodological accounts for the ontology of art. The next four sections are devoted to explore this issue.

4.3 Thomasson's Descriptivism

The distinction between descriptive and revisionary metaphysics is due to Peter Strawson, who presents the contrast between both views in the following terms: 'Descriptive metaphysics is content to describe the actual structure of our thought about the world; revisionary metaphysics is concerned to produce a better structure' (Strawson, 1959, 9). Thomasson endorses the descriptivist attitude in her main thesis in the metaontology of art. She argues that the discovery model is an inadequate explanation of the way in which we acquire knowledge about the ontology of art (Thomasson, 2005, 221). The discovery model is the idea that ontological inquiry may provide 'discoveries of fully determinate, mind-independent facts about the ontological status of works of art of various kinds, about which everyone may be ignorant or in error' (Thomasson, 2005, 221). Thomasson defends that, in contrast to discoveries of natural and biological kinds, the discovery model is mistakenly applied to the ontology of artworks due to the *qua problem* faced by causal theories of reference (Thomasson, 2005, 222–223). Those theories assume that natural kinds can be independently selected (e.g. gold) by ostensibly pointing to a sample of them. However, each sample includes entities of a very different nature (e.g. chemical, functional, biological, etc.). Determining the reference to any kind is not possible without an additional disambiguating concept that specifies the sort of kind pointed to. Crucially, Thomasson observes that the disambiguation of the reference of art-kind terms is different from the disambiguation of terms for natural kinds. Regarding natural kinds, the reference of the term is specified stipulating the sort of kind referred to. However, in Thomasson's view, the disambiguation of the reference of an art-kind term requires two conditions (Thomasson, 2005, 225; Thomasson, 2004, 86): [1] The grounder of the term's reference must think that the reference of that term will be an art kind; [2] she must have a previous background conception of the ontological status of the kind of artworks considered. According to [2], a pre-theoretical ontological conception is, in Thomasson's view, indispensable for grounding the reference of terms for art kinds, such as 'symphony', 'novel', 'poem' or 'painting' (Thomasson, 2006, 248). Art-kind terms, Thomasson argues, function as sortal terms, and to establish the reference of them, the grounder must have a general idea of the existence, persistence and identity conditions of the things that will be named by this term (Thomasson, 2010, 120).

Thomasson's argument against the discovery model in the ontology of art is that, since a background ontological conception is indispensable for referring to artworks, and we are successful in our referential practices, this background ontological conception is not subject to a massive error. Consequently, ontological discoveries entailing revisions of the background ontological conception are unjustified. There cannot be art-ontological discoveries because there is nothing to be discovered beyond the tacit ontological conception of our artistic practices (Thomasson, 2005, 227). In other words, Thomasson rejects the thesis of metaontological realism. First, we cannot refer directly to paintings, novels or symphonies apart from a pre-theoretical ontological conception. Second, the pre-theoretical conception implicit in our artistic practices determines the ontological status of the referents of art-kind terms (their identity, existence and persistence conditions). The use of art-kind terms qua sortal terms is ruled by application and co-application conditions (Thomasson, 2010, 120). Application conditions fix the existence conditions of *Fs* by setting the relevant conditions in which the application of the term 'F' is successful. In turn, co-application conditions fix the identity and persistence conditions of *Fs* by determining when 'F' may be reapplied to the same thing. Therefore, we cannot go in to investigate the nature of artworks and obtain surprising ontological discoveries.

The task of art ontology is, consequently, regarded as a descriptive one. It consists in making explicit this tacit ontological conception in a more formal and systematic way using the method of conceptual analysis (Thomasson, 2010, 120–1; Thomasson, 2006, 251–2; Thomasson, 2005, 226). This is the only right way to specify the ontological status of artworks. Thomasson is clear on this point:

> The *only* appropriate method for determining their ontological status is to attempt to unearth and make explicit the assumptions about ontological status built into the relevant practices and beliefs of those dealing with works of art, to systematize these, and to put them into philosophical terms so that we may assess their place in an overall ontological scheme (Thomasson, 2004, 87–88).

The indispensability of a tacit ontological conception to determine the reference of our art-terms implies, according to Thomasson, that artworks' nature is determined by that tacit conception. Accordingly, it is impossible to obtain ontological knowledge of artworks inconsistent with the ontological conception implicit in the system of beliefs that governs our artistic practices. Only ontologies that cohere with the tacit ontological conception are methodologically apt. The only legitimate task of the ontology of art is to make explicit the ontological conception implicit in artistic practices.

4.4 Thomasson's Descriptivism and the Problem of Triviality

In this section, I argue that Thomasson's descriptivism provides a methodology that does not ensure ontological accounts to achieve the ultimate goal of the ontology of art (for an alternative criticism, see Puy 2022). As noted in Sect. 4.2, this goal consists in developing ontological accounts that have epistemic value to increase our understanding of artworks and the practices to produce and properly appreciate them. Thomasson's methodology does not secure the achievement of this goal because it falls under a problem of *triviality*: as we will see, it unjustifiably reduces art ontology to a trivial task. To prove this charge, two things must be shown: first, that Thomasson's descriptivism reduces art ontology to a trivial task; and second, that this reduction is unjustified.

Starting with the first issue, the reduction of art ontology to a trivial task can be found in Thomasson's rejection of the discovery model. Thomasson's methodology prevents art ontology from supplying new knowledge about the nature of artworks. Under her methodology, art ontology is only entitled to make explicit what we *already knew* about them. For instance, if the tacit ontological conception characterizes paintings as things created at a certain time, spatially located and capable of surviving beyond their artists' life, these are the existence, identity and persistence conditions that a right ontological account must announce, and *nothing else*. However, this is information about the nature of paintings that we tacitly knew *before* doing ontology. Since ontology cannot supply new knowledge (i.e. surprising discoveries) about the nature of artworks, art ontology does not increases the understanding of artworks and the world of art. After all, previously to ontological inquiry, we were managing well enough with artworks, and our practices were reasonably successful, which reveals a sufficient degree of understanding of artworks and the practices that involve them. No additional understanding in a substantive sense would be supplied by ontological accounts that follow Thomasson's descriptivism. Accordingly, art ontology would not satisfy the ultimate goal of ontology in order to be a substantive task.

We might think, nonetheless, that this reduction of art ontology is *justified*, and hence, that we *should* give up doing art ontology, or less crudely, regard it as a *non-substantive* task. Recall that Thomasson arrives to descriptivism from an independent ground, the *qua problem*, i.e. from the need of a disambiguating concept for fixing the reference of an art-kind term. Thus, her methodological approach is independently grounded in a particular account of reference-fixing. However, even assuming that the reference of an art-kind term requires a minimal conceptual content to be fixed, Julian Dodd has shown that there is an alternative account on reference-fixing that does not entail the rejection of the discovery model. Following Gareth Evans (1985), Dodd argues that this conceptual content does not fix the reference of the term by fit—i.e. by determining the identity, existence and persistence conditions to be satisfied by the things referred by the term—, but by causal origin—i.e. the referent of the term is that thing, or that kind of things, that is the dominant or causal source of the relevant body of information that the grounders of the reference associate with the term (cf. Dodd, 2012, 84). Accordingly, it is possible for that content to contain inaccuracies regarding the identity, existence and persistence conditions of the thing, or kind of things, referred by the term (Dodd, 2012, 85).

It has been recently criticised to Thomasson's (2015) general metaontological framework, broadly known as 'easy ontology', that easy inferences are *neutral* with respect to the *nature* of properties (Bueno & Cumpa, 2020). As Otávio Bueno and Javier Cumpa have pointed out, the easy inference from 'The house is red' to 'There is a property, redness, that the house has' *might* commit us to the existence of the property of being red, but it does not commit us to any particular ontological thesis (platonism, nominalism, tropism, etc.) about the nature of such property. An analogous phenomenon arises for art-kind terms if we embrace Evan's theory of reference-fixing: the tacit ontological conception would be *neutral* regarding the *nature* of artworks. Evans proposes his theory to account for the phenomenon that a subject *S* may refer to a thing *x* by means of a description that *x* does not satisfy (cf. Evans, 1985: 14–6). In the majority of cases, according to Evans, *x* is the thing that is dominantly or causally responsible for the body of information that *S* possesses—the set of descriptions that *S* associates with the term with which *S* intends to refer to *x*. For *S* to refer to *x* by means of 'X', it is not necessary that *x* satisfies the set of descriptions that *S* associates with 'X'. The existence, persistence and identity conditions contained in the application and co-application conditions of art-kind terms are thus neutral with respect to artworks' nature insofar as they arise just to satisfy contingent pragmatic goals of our artistic practices and may be mistaken about artworks' nature. In addition, the satisfaction of the application and co- application conditions of a term is not *simpliciter*. It depends on the logic adopted to describe the thing referred by the term, with the bizarre consequence that the existence, persistence and identity conditions of that thing would depend on the logic adopted rather than in facts about that thing (cf. Bueno & Cumpa, 2020).

Therefore, Evans' theory entitles ontological inquiry to achieve discoveries about the nature of artworks that revise our tacit ontological conception while avoiding, at the same time, the *qua problem*. The neutralism of the tacit ontological conception enables art ontology to provide new knowledge about artworks' nature, and hence, to increase our understanding of them. It opens the door for art ontology to be substantive. So, why should we prefer Thomasson's account on reference-fixing rather than Evan's view? Thus, if the reduction of art ontology to a trivial task is based on a particular account on reference-fixing, and the choice of this account on reference- fixing is not justified, the trivial reduction of art ontology is *not justified*.

It might be argued nonetheless that the charge of trivial reduction is too uncharitable with Thomasson's descriptivism. Indeed, metaontological descriptivism can be understood in a *strong* or in a *weak* sense (Chalmers, 2009, 84–5). Both senses share the idea that claims concerning our folk ontological conception about *Fs* and formal ontological claims about *Fs* have the same correctness conditions given by our tacit folk ontology. However, only *strong descriptivism* regards the coincidence between folk and formal ontology as a trivial fact about the world. According to strong descriptivism, folk and formal ontological claims would have the same cognitive content. By contrast, *weak descriptivism* opens the possibility for art ontology to make a substantive contribution to understanding the nature of artworks and artistic practices. Weak descriptivism takes folk and formal ontological claims as differing in their cognitive content, despite coinciding in their correctness conditions. Consequently, making explicit the tacit ontological conception about artworks in a systematic and formal ontological account would increase our understanding of artworks and the world of art. If that were the case, art ontological accounts that follow Thomasson's methodology would satisfy the ultimate goal of ontology, and art ontology would be a substantive enterprise.

Let us assume for the sake of the argument that Thomasson's descriptivism is a weak form of descriptivism. However, even in this case, her methodology does not escape from the problem of triviality. Art ontology is not usually appealed to when our artistic practices work well and are not problematic. Instead, we typically ask for ontology in cases of *conflicting practices*, i.e. local domains of our artistic practices in which our tacit ontological conception is ambiguous or confusing. These are areas of artistic practices in which we do not have a clear tacit knowledge about the entities involved. What is expected from ontology, and what in fact it does, is a substantive contribution to our understanding of those items and practices at stake, *even* if this contribution entails the revision of some beliefs of the tacit ontological conception. The problem is that Thomasson's descriptivism offers no methodological tools for specifying the ontological status of artworks involved in those cases of conflicting practices. Let me expand on these points.

Ontology substantively contributes to improve and accurate our aesthetic appreciation of artworks in cases of conflicting practices. As Dominic Lopes points out, the activities involved in a correct appreciation of an artistic object depend on its ontological nature (cf. Lopes, 2007, 243). Some objects are to be looked at, others to be heard, others to be heard or looked at in themselves, but others in relation to other objects, and so on. In addition, ontology does not only help us to know the activities to be done to rightly appreciate an artistic object. Art forms are medium-dependent, i.e., they depend on the kind of stuff-manipulations by an artist to generate the artistic object (Lopes, 2007, 247; Dodd, 2016, 257). Artworks belonging to the same art form are expected to belong to the same ontological category, and this fact generates expectations in the audience about what to look for when appreciating an artistic object. When the tacit ontological conception about artworks is ambiguous or confusing, the appreciation of the artworks involved turns out to be problematic and sometimes bound up to a systematic failure. Ontological inquiry is extremely welcome in those cases and it helps us to place artistic objects in its right art form, which contributes to generate the right expectations when it comes to appreciate them. Brief examples can illustrate this phenomenon.

John Cage *4′ 33″* has received highly discrepant appreciative judgments in our artistic practices. It has been characterized as 'absolutely ridiculous' or 'stupid', as well as 'sublime' and 'original' (cf. Dodd, 2018, 631). The source of these contrasting judgments is that the tacit ontological intuitions about what kind of thing is *4′33″* are confusing. Some people regard each performance of *4′33″* as the performance of a new piece (Gann, 2010, 185), others as a piece of environmental site-specific sounds (Tomkins, 1968, 115), others as a silent piece of music, and so on.

Recent research in the ontology of music has clarified the ontological status of *4′ 33″* as a type of action instead of a type of sound-sequence events (Dodd, 2018, 633–636), which qualifies it as a work that falls under the art form of performance art rather than music art (Dodd, 2018; cf. Davies, 2003). Ontology has contributed to this extent to put in right place the sort of expectations with which the audience should approach this work to adequately appreciate it. Ontology has done this supplying understanding about this work's nature and revising the widespread intuition that each artwork specified in a musical score is a work of music, and hence a type of sound-sequence events.

Conflicting practices do not only attain individual artworks but also art kinds. This is the case of *readymades*. General audiences usually deal with them as if they were sculptures, under the tacit ontological assumption that they are concrete particulars having the existence, persistence and identity conditions of sculptures. The way sometimes readymades are exhibited in museums reinforces the idea that they are plastic art, and more specifically, sculptures. However, engaging with readymades as if they were sculptures leads people to a systematic appreciative failure: they feel 'frustrated, confuse and dismissive' with those works (Lopes, 2007, 238). The reason is that, appreciated under the category of sculptures, readymades do not satisfy people's expectations about the medium of those works, i.e. about the kind of stuff- manipulations that artists should make to generate those works. With the *performance theory*, ontology has provided ontological discoveries about the nature of those artworks that may help to avoid this appreciative failure: readymades are action tokens (events) and the artistic object is only the focus of appreciation of the work (cf. Davies, 2004). Consequently, the work qua object of artistic and aesthetic appreciation is not the artistic object, but the whole event or action generated by the artist to specify that object as a focus of appreciation. Once again, ontology has contributed to put in right place people's expectations for artistic and aesthetic appreciation by means of supplying understanding about the nature of a particular kind of artworks.

But conflicting practices do not only concern post-avant-garde works as, for instance, the case of musical versions illustrates. Musical versions are intentional: they are versions *of* a work. For instance, tribute bands make versions *of* The Beatles' songs, and Tchaikovsky composed two versions *of* his *Second Symphony*. But what is the work? In other terms, what is the final object of aesthetic appreciation, and what is the appreciative role of the intermediate objects (i.e. the work's versions and the performances of those versions)? It might be thought that the work is the sound structure indicated by the composer's original score or performance. Thus, the sound structure of the performance of *Yellow Submarine* recorded in 1966 by The Beatles would be the intentional object of aesthetic appreciation of a version made by a tribute band. But what happens then when Tchaikovsky decided to compose a new version of his *Second Symphony* because he was not satisfied with the original one? Should we identify the work with the final and improved version, the last one approved by the composer? Was the work *completed* with this final version? Or, rather, did his *Second Symphony* evolve through time, having thus different *stages*? But, if versions' sound structures are slightly different, why not to regard them as different but similar works? However, in that case, what would be the difference between versions and inspired works, or works borrowing from a source work? The intuitions involved in the tacit ontological conception about versions, and derivatively about musical works, are thus confusing, and the objects of aesthetic appreciation are not clearly identified. However, recent research in the ontology of music has shed light on this point, offering a picture in which versions are repeatable entities (types of lower order) that are instances of a musical work (a type of higher order individuated by a less specific sound structure) (cf. Puy, 2019, 247–8). Musical versions are to be appreciated as different ways of presenting a same work by means of offering different ways of performing it. The *nested types theory* has contributed this way to the appreciation of those musical items by means of revising the intuition that a musical work is individuated by the sound structure indicated by a score.

Crucially, these cases of conflicting practices are problematic for Thomasson. In Thomasson's account, if the tacit ontological conception is ambiguous, it cannot select the sort of thing that we are referring to with an art-kind term, like'musical work', 'version' or 'readymade'. If so, there are no determined answers to the questions about the identity, existence and persistence conditions of the things referred to by those terms. Consequently, art ontology would be inhibited from offering a substantive contribution in cases of conflicting practices, scenarios in which ontological knowledge is particularly welcome due to our lack of understanding within those domains. Our interest in doing ontology is not to know what we tacitly think about entities of a particular domain, but rather to know the nature of those entities. As John Heil has observed concerning scientific and ordinary domains, our interest is not to know our ordinary and scientific claims that are true, but rather to know the nature of the things that make scientific and ordinary claims true (cf. Heil 2012, 2, 4–5). Similarly, in the artistic domain, David Davies has argued that ontologies of art are not about the meanings of art-kind terms, but about 'the nature of the object of critical and appreciative attention' (Davies, 2017, 126). And what I have precisely highlighted here is that what motivates that interest in ontology is an interest in *understanding*: an interest in better understanding the nature of artworks that may help us to solve the problems we face in conflicting practices, improving the way of engaging with and appreciating the artworks involved in those domains. Art ontology is substantive only if it satisfies this interest.

Therefore, ontological inquiry would be trivial if we follow Thomasson's methodology. If we regard it as a weak descriptivism, ontological accounts that satisfy her methodology are non-trivial in domains where we already have a sufficient understanding of artworks and practices. However, in domains where we are in need of understanding artworks' nature (conflicting practices), Thomasson's methodology entails that art ontology must remain silent. But, as the previous examples illustrate, ontology really makes substantive contributions in those domains. Thomasson's methodology does not provide us with tools to validate and explain this substantive contribution that ontology can and, in fact, makes. Instead, grounded on a particular view of reference-fixing arbitrarily chosen, it unjustifiably prevents those contributions that usually constitute departures from, and not mere descriptions of, the tacit ontological conception in order to improve our appreciative practices. In last term, Thomasson's methodology is *revisionary* of *our practice of exchanging knowledge* between the artworld and ontology.

4.5 Dodd's Revisionism

In contrast to descriptivism, metaontological revisionism regards ontology as an instrument of conceptual change to produce a better structure of thought about the world. To this extent, it pursues much clearly than descriptivism the ultimate goal of ontology of increasing our understanding of the world. Revisionism stresses that our ordinary concepts have a pragmatic function: they play a fundamental role in making our ordinary practices successful regarding a broad variety of purposes. In the musical domain, for instance, they play a fundamental role in communication between musicians in a rehearsal, in a composer's explanation of her work, in people expressing their preferences for Beethoven's *5th Symphony* over Schubert *5th Symphony*, and so on. Ordinary concepts arise and are articulated to play a practical function. However, they have not been designed to play a theoretical or epistemic function. Their goal is not *explanation*. Accordingly, the revisionist argues that they lack precision to offer a secure way to determine the nature of artworks.

Dodd offers one of the most sophisticated defences of metaontological revisionism in the artistic domain. He states his view claiming that 'the correct ontological theory of a given artwork kind might be substantially revisionary of our folk ontological conception of things of this kind' (Dodd, 2013, 1048). The defence of this viewpoint requires the assumption the thesis of *folk-theoretic modesty*, according to which our ordinary ontological conception of artworks 'might be substantially mistaken' (Dodd, 2013, 1048). In other terms, revisionism entails an error theory for our pre- theoretical ontological conception of artworks. However, this error theory is only possible if the nature of artworks is not determined by the folk ontological conception implicit in our artistic practices. Accordingly, metaontological revisionism assumes the thesis of metaontological realism, namely, that the correct answers to the fundamental questions of ontology are mind-independent and objective (cf. Dodd, 2013, 1048).

If the nature of *Fs* is mind-independent, the door is open to the discovery model. Conceptual analysis is thus not regarded by revisionism as a suitable method to determine the nature of *Fs*. Alternatively, Dodd argues that the proper methodology to investigate the nature of *Fs* is akin to a quasi-scientific methodology. The answers about the nature of *Fs* are regarded 'as hypotheses that are assessed according to a loose battery of criteria for choice', among which are 'explanatory power, simplicity and integration with the findings of other domains', together with the coherence with the folk thought about *Fs* (Dodd, 2013, 1051). Therefore, contrary to Thomasson's view, Dodd's revisionism holds that the criterion to decide between different candidates about the ontological nature of *Fs* is not merely the coherence with our pre-theoretical ontological conception about *Fs*. Since a balance has to be made among different criteria, the theory selected can be more or less revisionary of our folk thought about *Fs*.

As noted in the previous section, even if we assume that the reference of an art-kind term requires a minimal conceptual content to be fixed, Dodd argues that Evans' theory of reference-fixing enables the discovery model in art ontology. The discovery model, grounded in metaontological realism, is regarded by Dodd as the default position in the metaontology of art, taking as examples the views offered by Gregory Currie, Mark Sagoff, Nicholas Wolterstorff and Peter Kivy (cf. Dodd, 2012, 77). There are two facts that, according to Dodd, favour this phenomenon. First, ontological proposals about *Fs* are in the 'explanation business': their main goal is to offer the best explanation about the nature of *Fs* (Dodd, 2012, 77). The best ontological theory is not the one that best coheres with our folk ontological conception, but the one that offers the best explanation of *Fs* in light of theoretical virtues. Accordingly, the best ontological theory may be one that revises some of our folk beliefs about the nature of *Fs*. Second, there is a disanalogy between an ontological proposal about *Fs* and our folk ontological conception about *Fs*. Ontological proposals about *Fs* are constrained by the need of explaining the phenomena about *Fs* that raise relevant philosophical questions and puzzles. By contrast, the folk ontology about *Fs* lacks this constraint because its goal is not explanation, but rather to offer a concept of *Fs* that plays a successful role in our ordinary practices involving *Fs* (Dodd, 2012, 79). The configuration of ordinary concepts about artworks is thus guided by targets others than those that guide ontological inquiry. Since the target of folk concepts about artworks is not explanatory but pragmatic, they might cover up mistakes about their nature.

4.6 Dodd's Revisionism and the Problem of Solipsism

In comparison with Thomasson's descriptivism, Dodd's revisionism has the advantage of methodologically validating the substantive task of art-ontological accounts in the cases of conflicting practices introduced in Sect. 4.4. However, the sense of 'explanation' stressed by Dodd seems to be motivated by a sense of 'epistemic value' akin to the one defended by Sider in joint-carving terms. As noted in Sect. 4.2, this sense of 'epistemic value' is too narrow, and hence inadequate, to achieve the holistic understanding of the world's images pursued by general ontology, and the understanding of the artworld and its objects pursued by art ontology. The endorsement of Sider's notion of 'epistemic value' leads Dodd's methodology to provide no clear limits to the discovery model in the ontology of art. To this extent, it supplies no articulated way of constraining the *kind* of beliefs of our folk ontological conception that may be revised by art-ontological inquiry. As a result, Dodd's revisionism faces a *problem of solipsism*: it is a methodology that does not guarantee conceptual mediation between the outcomes of ontological inquiry and the ordinary view of our artistic practices. Thinking and speaking solely in joint-carving terms, as suggested by Sider's criterion, does not help ontology to supply understanding of artworks and the artistic productive and appreciative practices they involve.

It is true that the Dodd's revisionism claims for a balance between different theoretical virtues, and that 'this set of virtues includes coherence with folk belief' (Dodd, 2013, 1051). However, he provides neither tools nor criteria to obtain this balance, and the door is open to interpret the balance as a quantitative matter. The crucial point is that the coherence with folk belief is rather a *qualitative* matter. Not all folk beliefs are the same. Some of them are fundamental to our artistic practices, and entail many other beliefs. The revision of a fundamental belief may imply that our artistic practices are mistaken as a whole. For instance, an ontological account that concludes that there are no musical works is more revisionary than one that concludes that musical works are not creatable things and that their identity is not tied to their composers and instrumentation. The latter account *prima facie* revises three beliefs, while the former just one. However, the belief revised by the former is more fundamental to our musical practices. A revision of this belief would entail that most of our musical practices are a *simulacrum*. A composer would not really produce *something* called musical work, managers commissioning compositions would not be really paying for a commodity, concerts announced as premieres of works would be premieres of *nothing*, listeners interested in hearing and appreciating Beethoven's *5th Symphony* would not be really interested in appreciating an aesthetic object, a performer's attitude of providing a new interpretation and insight into a work would be foolish, and so on. Insofar Dodd's revisionism does not discriminate between kinds of intuitions that may be revised, it puts no clear constraints to the discovery model. The revision of one fundamental belief may entail that our musical practices are mistaken as a whole, whereas the same consequence is not entailed by the revision of three less fundamental beliefs.

Since Dodd's revisionism provides no constraints for the kind of folk intuitions to be revised, an ontological account that concludes that there are no musical works can be methodologically valid in case it satisfies the theoretical virtues of simplicity, explanatory power and integration with the findings of other domains. To this extent, Dodd's revisionism is open to a problem of solipsism. A revision of certain intuitions of our musical practices may lead to change the *subject matter* about the concept of musical work, version, performance, or any object concerned by art ontological inquiry. It is thus a methodology that may validate ontologies whose outcomes are not translatable to our artistic practices, breaking the bridges between them. It runs the risk of confining art ontology to a philosophers' game, being unable to make relevant contributions to our artistic practices. Concerning the philosophy and metaphysics of science, some authors have warned that 'when philosophy becomes institutionally solipsistic, it risks making itself intellectually and culturally irrelevant' (Ladyman & Ross, 2007, 26). The same applies in the domain of the ontology of art.

This solipsistic risk is twofold problematic for Dodd's revisionism. First, it is a methodology that does not guarantee that ontological accounts may achieve the main motivation of revisionism: the improvement of the conceptual structure that grounds our artistic practices. Derivatively, it is a methodology that does not secure the ultimate goal of art ontology: it does not guarantee that ontological accounts may offer a substantive epistemic contribution to our artistic practices in the sense pointed in Sect. 4.2. Since revisionism does not guarantee conceptual mediation between the outcomes of ontological inquiry and our artistic practices, it may promote accounts that are interesting for the ontologist but irrelevant (i.e. without epistemic value) for our practices. From a different perspective than Thomasson's descriptivism, ontological accounts fitting Dodd's revisionism may fail to be substantive for our artistic practices.

In addition, a *global* mistake of our folk concepts about the nature of artworks is not plausible given the reasonable success of our artistic practices as a whole. If the tacit ontological conception of our artistic practices were radically mistaken, such practices would not be successful to achieve their main pragmatic goals. But they are successful to a reasonable extent, so no motivations would be found for a global revision. By contrast, as noted in Sect. 4.4, a substantial contribution of ontology takes place in *local* domains of our practices: the cases of conflicting practices. It is there where our artistic practices are confusing and inefficient to achieve their targets. It is there where the agents of artistic practices demand and are receptive of solutions coming from ontology.

4.7 A Goodmanian Solution

As indicated in Sect. 4.2, the ultimate desideratum to be pursued by a methodological account in the ontology of art is to ensure epistemic value for understanding artworks and the productive and appreciative practices they involve. The problems of triviality and solipsism suggest that the achievement of this goal implies two complementary methodological principles. To avoid the problem of triviality without falling under the problem of solipsism, a methodological account must allow local revisions of our folk ontological conception about artworks. This principle, which I will call 'minimal revisionism', can be stated as follows:

> Ontological accounts of artworks should be able to revise our folk intuitions about artworks whenever they clash with sound theoretical principles.

In turn, to avoid the problem of solipsism without falling under the problem of triviality, a methodological account must preserve a minimal descriptivist attitude towards our artistic practices. This second principle, which I will call 'minimal descriptivism', can be formulated as follows:

> Ontological accounts of artworks, within a given context, should be able to accommodate widespread practical intuitions about art.

The principles of minimal revisionism and minimal descriptivism imply a mutual adjustment between artistic practices and art ontology. But how could this adjustment be achieved?

It was noted in the previous section that Dodd's revisionism is in a better position than Thomasson's descriptivism to satisfy both principles. Indeed, his account claims for a balance between different theoretical virtues, including the coherence with folk intuitions about artworks' nature. The problem of solipsism appears insofar his account does not specify criteria or tools to achieve this balance. In last term, the problem was located in the inability of Dodd's revisionism to identify the kind of folk intuitions that can be revised in order to secure conceptual mediation with our artistic practices.

Goodman's distinction between the notions of 'entrenchment' and 'familiarity' of predicates provides a promising methodological tool to identify the kind of intuitions that can be revised. According to Goodman, the entrenchment of a predicate in our linguistic practices is relative to the actual projections of that predicate and of all the predicates coextensive with it (Goodman, 1964/1983, 95). A predicate is actually projected in a hypothesis if the hypothesis 'is adopted after some of its instances have been examined and determined to be true, and before the rest have been examined' (Goodman, 1964/1983, 87). In other words, the predicate is actually projected if the hypothesis is supported—it has positive (true) instances–, unviolated—it has no negative (false) instances–, unexhausted—it has undetermined cases (neither positive nor false), mainly future cases–, and if all the hypotheses conflicting with it have been overridden (Goodman, 1964/1983, 90, 104). Consequently, the more a predicate is actually projected, the more entrenched it is.

The entrenchment of a predicate derives thus from linguistic use. However, Goodman is very careful at this point distinguishing between entrenchment and familiarity:

> Entrenchment and familiarity are not the same. An entirely unfamiliar predicate may be very well entrenched (…) if predicates coextensive with it have often been projected (…). Again, a very familiar predicate may be rather poorly entrenched, since entrenchment depends upon frequency of projection rather than upon mere frequency of use (Goodman, 1964/1983, 90).

This distinction between entrenchment and familiarity can be straightforwardly applied to linguistic practices concerning artworks. A specific use of a predicate might occur with high frequency in our ordinary artistic discourse and being, nonetheless, a non-entrenched use of that predicate because it has occurrences only in non-projectible hypotheses, i.e. hypotheses that are either non-supported (with no positive instances), or violated (with negative instances), or exhausted (with no undetermined cases) or ones that conflict with non-overridden hypotheses. These considerations allow us to distinguish between familiar and entrenched intuitions. *Familiar intuitions* are those that, when made explicit, are widely shared and appealed to in our artistic practices for different practical purposes. By contrast, *entrenched intuitions* are those that, in addition to be familiar, also are projectible hypotheses when made explicit, i.e. ones that are supported, non-violated, non-exhausted and that do not conflict with non-overridden hypotheses. Entrenched intuitions, for their characteristics, play the role of constituting the core of our artistic practices. They secure the success of our artistic practices and make them compatible with other everyday and scientific practices that are necessary for, and complementary to, artistic practices (from going shopping to calibrating different alloys for brass instruments).

In light of the distinction between entrenched and familiar intuitions, an important remark is to be made. Whereas we can determine what intuitions are familiar to our artistic practices, we cannot positively determine which intuitions are entrenched. Familiar intuitions are widespread in our practices. It is a logical work to detect the ontological presuppositions that ground usual claims recurrently made in our artistic practices. In addition, frequency of use is a quantitative matter. Consequently, familiar intuitions can be detected and measured by means of empirical methods. Testing the intuitions of people concerning a case of ontological relevance by means of vignettes, or confronting them with visual or musical stimuli in an experiment, as well as empirical research in linguistic corpora, are ways to determine whether an intuition is familiar to our artistic practices. Although the use of empirical methods is not free of problems (cf. Machery, 2017; Zangwill, 2019), it is a way of ascertaining familiar intuitions that secures a minimum of independency of the philosopher's prejudices (cf. Bartel, 2018). In turn, the entrenched status of an intuition is not something that can be positively shown. Instead, we can only show whether an intuition is *not* entrenched in our artistic practices by means of showing if it is a hypothesis that has no positive cases, or has negative cases, or has no indeterminate cases, or if it conflicts with other non-overridden hypotheses or sound theoretical principles. In other words, it is a matter of showing that the intuition is not a projectible hypothesis, something that is done by means of philosophical (ontological) examination.

Consequently, the ontologist's work with respect to the intuitions of our artistic practices involves two steps. Firstly, she has to identify the set of familiar intuitions by means of logical analysis and empirical methods. The ontologist must start with a close examination of our artistic discourse, preserving this way part of the descriptivist spirit in the metaontology of art. Secondly, she has to eliminate, as a result of an analytical process, those familiar intuitions that clash with non-overridden hypotheses or sound theoretical principles. Every familiar intuition must be accommodated by art ontological accounts, unless it's not entrenched, but a merely familiar one. This constraint secures a conceptual mediation between ontology and artistic practices. Art ontology is methodologically entitled to revise only *familiar non-entrenched intuitions* if there are good reasons for such a revision. Non-entrenched intuitions are a symptom of cases of conflicting practices, in which opposed beliefs are on a same plane to guide artistic practices. As a result, artistic practices might have not a clear guidance in those particular areas, and it is there where methodologically valid art ontological accounts should have an epistemically valuable contribution. The distinction between familiar and entrenched intuitions provides us with a way of satisfying the principles of minimal revisionism and minimal descriptivism and to achieve the ultimate goal of the ontology of art.

The role of traditional theoretical virtues—such as explanatory power, simplicity or the compatibility with our best theories in other domains—is now integrated and put at the service of the identification of the intuitions that are not entrenched. The virtue of compatibility with our best theories in other domains helps us to elucidate whether an intuition clashes with non-overridden hypotheses. In turn, explanatory power and simplicity helps to decide between two ontological candidates that accommodate equally well the familiar ontological intuitions that are not revealed as non-entrenched ones.

Let me finish with an illustrative example of how these methodological tools may work. Let us suppose that by means of empirical methods we determine that the following pattern is usually instantiated in our musical discourse: *x* revised *w* by means of modifying *z* (where *x* is a variable for the name of a composer, *w* is a variable for the name of a musical work, and *z* a variable for musical items such as chords, melodies, tonalities or instrumentation). Then, let us suppose that by means of logical analysis, we identify that the use of this pattern is based on the following presupposition: musical works can change over time, i.e. musical works are temporally flexible entities (cf. Rohrbaugh, 2003). However, by means of empirical and philosophical analysis, we observe that those works are performable (and sometimes *performed*) in the previous states to being revised, and hence that the hypothesis that musical works are temporally flexible conflicts with the non-overridden hypothesis that if an object changes, it no longer exists in its previous state (cf. Dodd 2007, 88–9). The non-overridden hypothesis of change is entrenched insofar as it is highly projected. It is presupposed by a high number of other projectible hypotheses in any field. For instance, concerning the growth of babies, we expect a baby to be bigger when she is 8 months old than when she was 2 months old. We would say that she has changed because we would not expect that she is now in the same state as when she was 2 months old. Or, in physics, when we say that some water changes from solid to liquid state, we say that it no longer is in its previous state. Therefore, the hypothesis that musical works are temporally flexible is not projectible because it conflicts with the non-overridden hypothesis that if an object changes, it no longer exists in its previous state. Art ontology is entitled thus to revise this non-entrenched intuition, regardless its familiarity in our musical discourse.

4.8 Conclusions

In this paper, I have argued that Thomasson's descriptivism and Dodd's revisionism do not provide adequate methodologies for the ontology of art. They do not secure ontological accounts to have epistemic value for understanding artworks and the artistic practices they involve. As stated in Sect. 4.2, this is the ultimate goal of the ontology of art if it aims to be a substantive, and hence, a valuable task. I have argued that Thomasson's descriptivism and Dodd's revisionism do not ensure the achievement of this goal because they face two problems, respectively.

Thomasson's methodology faces a problem of triviality, unjustifiably reducing art ontology to a trivial task. First, her methodology is reductive of the epistemic scope of art ontology. Her rejection of the discovery model prevents art ontological accounts from supplying new knowledge about the nature of artworks beyond the knowledge we implicitly have in our artistic practices. Second, this reduction of art ontology has been revealed as unjustified insofar as it is grounded in an arbitrary choice of a reference-fixing theory for art-kind terms. Third, in a charitable approach to Thomasson's account, her methodology might be regarded as epistemically non-reductive of art ontology. However, four, this only applies to domains in which artistic practices are not problematic, but not to conflicting practices. Five, in cases of conflicting practices, art ontological accounts must remain silent according to Thomasson's methodology, since there would not be determinate answers to the questions about the existence, persistence and identity conditions of the works involved. However, six, conflicting practices are domains where people demand an understanding of artworks' nature that art ontology can supply and *in fact* provides. Therefore, seven, Thomasson's methodology cannot account for this fact, and hence, it unjustifiably reduces art ontology to a trivial enterprise.

In turn, Dodd's metaontology, as a revisionary account, focuses art ontology on improving our structure of thought about artworks. To this extent, it has been revealed as a methodological view that is in a better position than Thomasson's to achieve the ultimate goal of art ontology. However, it has been illustrated that Dodd's revisionism faces a problem of solipsism: it is a methodology that does not secure conceptual mediation between ontology and artistic practices. To this extent, Dodd's revisionism is not able ensure the ultimate aim of revisionism and art ontology. This problem has been located in the inability of Dodd's revisionism to identify the kind of intuitions that can be revised.

Goodman's distinction between familiarity and entrenchment has emerged here as a useful tool to identify those intuitions. Familiar intuitions are those that, when made explicit, are widely shared and appealed to in our artistic practices for different practical purposes. The identification of them is made by logical and empirical analysis. However, entrenched intuitions are those that, in addition to be familiar, also are projectible hypotheses when made explicit. They constitute the core of artistic practices. They cannot be positively identified. Instead, we can only ascertain which intuitions are not entrenched by means of philosophical analysis. Every familiar intuition must be accommodated by art ontological accounts, unless it is not entrenched. Art ontology is methodologically entitled to revise only *familiar non-entrenched intuitions* if there are good reasons demanded by the theoretical virtues of explanatory power, simplicity or integration with findings in other domains.

Acknowledgments This publication is part of the grant RYC2021-032014-I, funded by MCIN/AEI/10.13039/501100011033 and the European Union «NextGenerationEU»/PRTR, and of the project PID2023-147445NA-I00, funded by MCIN/AEI/10.13039/501100011033/FEDER, UE. An earlier version of this paper was given at the XXVIII Inter-University Workshop on Philosophy and Cognitive Science or the Sociedad Española de Filosofía Analítica (University of Granada, 2019). I sincerely thank the participants in those events, as well as Mirco Sambrotta and Javier Cumpa, for their helpful comments and suggestions.

References

Bartel, C. (2018). The ontology of musical works and the role of intuitions: An experimental study. *European Journal of Philosophy, 26*, 348–367.
Bueno, O., & Cumpa, J. (2020). Resisting easy inferences. *Philosophy and Phenomenological Research*. Forthcoming.
Buonomo, V. (2017). The scientific turn in metaphysics: A factualist approach. *Synthese, 198*, 793.
Chalmers, D. J. (2009). Ontological Anti-Realism. In D. J. Chalmers, D. Manley, & R. Wasserman (Eds.), *Metametaphysics* (pp. 77–129). Oxford University Press.
Cumpa, J. (2014). A materialist criterion of fundamentality. *American Philosophical Quarterly, 51*(4), 319–324.
Cumpa, J. (2018). Factualism and the scientific image. *International Journal of Philosophical Studies, 28*(5), 669–678.
Cumpa, J. (2020). Naturalism and the question of ontology. *American Philosophical Quarterly*. Forthcoming.
Davies, D. (2004). *Art as performance*. Blackwell Publishing.
Davies, D. (2017). Descriptivism and its discontents. *Journal of Aesthetics and Art Criticism, 75*(2), 117–129.
Davies, S. (2003). *Themes in the philosophy of music*. Oxford University Press.
Dodd, J. (2007). *Works of music: An essay in ontology*. Oxford University Press.
Dodd, J. (2012). Defending the discovery model in the ontology of art: A reply to Amie Thomasson on the qua problem. *British Journal of Aesthetics, 52*(1), 75–95.
Dodd, J. (2013). Adventures in the metaontology of art: Local descriptivism, artefacts and dreamcatchers. *Philosophical Studies, 165*, 1047–1068.
Dodd, J. (2016). The ontology of conceptual art: Against the *idea*. In J. Dodd (Ed.), *Art, mind and narrative. Themes from the work of Peter Goldie* (pp. 241–260). Oxford University Press.
Dodd, J. (2018). What *4′33″* is. *Australasian Journal of Philosophy, 96*(4), 629–641.
Evans, G. (1985). The causal theory of names. In G. Evans (Ed.), *Collected papers* (pp. 1–24). Clarendon Press.
Gann, K. (2010). *No such thing as silence: John cage's 4′33″*. Yale University Press.
Goodman, N. (1964/1983). Fact, fiction and forecast. Harvard University Press.
Heil, J. (2012). *The universe as we find it*. Oxford University Press.
Ladyman, J., & Ross, D. (2007). *Every thing must go. Metaphysics naturalized*. Oxford University Press.
Lopes, D. (2007). Conceptual art is not what it seems. In P. Goldie & E. Schellekens (Eds.), *Philosophy and conceptual art* (pp. 238–256). Oxford University Press.
Machery, E. (2017). *Philosophy within its proper bounds*. Oxford University Press.
Puy, N. G.-C. (2019). The ontology of musical versions: Introducing the hypothesis of nested types. *Journal of Aesthetics and Art Criticism, 77*(3), 241–254.
Puy, N. G.-C. (2021). Eliminating Martin's substratum-trope categorial ontology. *Synthese, 199*(5-6), 13009–13033. https://doi.org/10.1007/s11229-021-03364-5
Puy, N. G.-C. (2022). Descriptivism and the determination thesis: an untenable marriage in the metaontology of art abstract. *The Philosophical Quarterly, 72*(3), 595–614. https://doi.org/10.1093/pq/pqab055
Rohrbaugh, G. (2003). Artworks as historical individuals. *European Journal of Philosophy, 11*(2), 177–205. https://doi.org/10.1111/ejop.2003.11.issue-2. https://doi.org/10.1111/1468-0378.00182
Sellars, W. (1963/1991). Philosophy and the scientific image of man'. In Sellars, W. (Ed.), *Science, perception and reality* (pp. 1–40). Routledge and Kegan Paul.
Sider, T. (2011). *Writing the book of the world*. Oxford University Press.
Strawson, P. (1959). *Individuals: An essay in descriptive metaphysics*. Methue.
Thomasson, A. (2004). The ontology of art. In P. Kivy (Ed.), *The Blackwell guide to aesthetics* (pp. 78–92). Blackwell Publishing.

Thomasson, A. (2005). The ontology of art and knowledge in aesthetics. *Journal of Aesthetics and Art Criticism, 63*(3), 221–229.
Thomasson, A. (2006). Debates about the ontology of art: What are we doing here? *Philosophy Compass, 1*(3), 245–255.
Thomasson, A. (2010). Ontological Innovation in Art. *Journal of Aesthetics and Art Criticism, 68*(2), 119–130.
Thomasson, A. (2015). *Ontology made easy*. Oxford University Press.
Tomkins, C. (1968). *Ahead of the game*. Penguin.
Zangwill, N. (2019). Beauty and the agreeable: A critique of experimental aesthetics. In F. Cova & S. Réhault (Eds.), *Advances in experimental philosophy of aesthetics* (pp. 289–307). Bloomsbury Academic.

Chapter 5
A Deflationary Artifactualist Approach to the Ontology of Science

Xavier de Donato-Rodríguez and José L. Falguera

5.1 Introduction

In this paper, we aim to achieve a double goal. First, we intend to relate Thomasson's artifactualism to her deflationary approach to ontology. Then, we seek to explore the artifactualist option in detail as a serious way to understand scientific theories and models, and specifically the intentional direct referents of their terms: mathematical, theoretical and observational. Our proposal involves understanding the ontological commitment of science in a very peculiar way that is inspired in Carnap's distinction between external and internal questions (Carnap, 1950b), as well as in both Thomasson's easy approach to ontology (Thomasson, 2015) and her notion of an abstract artifact (Thomasson, 1999).

We will defend the idea that the ontological commitments that one has to adopt once one accepts a (well-confirmed) scientific theory are (or should be) only internal (in Carnap's sense). Strictly speaking, one should remain agnostic regarding the real existence of the entities that are posited by a given linguistic framework (in the sense of a hard ontologist). In this regard, we endorse a form of meta-ontological deflationism, in accordance with which such ontological commitments cannot be understood as affirming anything that a hard metaphysician might. So, when we accept a scientific theory, we use theoretical and observational terms and quantify over variables related to theoretical and observational entities. We also use mathematical expressions. However, our commitment to the existence of the referents of all these terms can only be internal.

Carnap (1956, 1958, 1966, 2000) also adopted a peculiar point of view about the referents of theoretical terms: he assumed that they are mathematical entities (and therefore, abstract objects). For reasons that we hope will become clear as this work

X. de Donato-Rodríguez (✉) · J. L. Falguera
University of Santiago de Compostela, Santiago, Spain
e-mail: xavier.dedonato@usc.es; joseluis.falguera@usc.es

M. Sambrotta (ed.), *Metaphysics Today: In Conversation with Amie Thomasson*, Synthese Library 512, https://doi.org/10.1007/978-3-031-98468-6_5

progresses, we reject Carnap's ultimate justification for considering the referents of theoretical terms to be mathematical entities; although our position is that Carnap was right insofar as he considered them to be abstract. More particularly, they can be seen as abstract *artifacts* in the sense of Thomasson (1999). In fact, we think that in some relevant sense, they are like the fictional characters of novels or stories. Although both theoretical entities and fictional characters are kinds of abstract artifacts in this sense, there are clearly important differences between them. While theoretical terms are considered (within theories) in order to explain and predict observational facts, fictional characters are considered (in novels or stories, for example) mainly to entertain. So, we may affirm the existence of theoretical entities, as abstract artifacts, but just internally within the linguistic framework of a theory (or the linguistic formulation of a scientific model).

Although it may not be such a novel idea to consider models (and theories themselves) as abstract objects (artifacts), the thesis that the theoretical entities that intervene in those models and theories are abstract artifacts too is much less frequent. We nevertheless maintain that this can be perfectly considered a Carnapian idea. Moreover, it fits perfectly well with Thomasson's artifactualism regarding fictional characters, and cultural and institutional products in general. We also intend to show how this idea fits well (in some sense) with Zalta's abstract objects theory (in Sect. 5.2), though we point out some particularities of that theory and compare them with those of Thomasson's. Our conclusion will be that, leaving some difficulties aside (difficulties that will be discussed), Thomasson's easy ontology approach combined with her artifactualism provides an adequate framework for an understanding of the ontology for science; this is something that may have remained unnoticed by the author herself (see, nevertheless, Thomasson (2020).

The structure of the rest of this paper is as follows. In Sect. 5.2, we focus on the issue of how fiction and scientific discourse are related. We defend the plausibility of applying the same type of ontology to models, theories and theoretical entities, and of understanding them as kinds of abstract artifacts. We do this by considering two well-known theories of abstract objects: Zalta's axiomatic theory and Thomasson's artifactualism. In Sect. 5.3, we present the central ideas of a Carnapian approach to ontology and meta-ontology. Within this, in Sect. 5.3.1, we discuss Thomasson's easy approach; and in Sect. 5.3.2, we present Carnap's original ideas. One of our main aims is to defend those ideas from misunderstandings that seem to have been derived from attributing to Carnap certain theses that we cannot believe he would accept. In this regard, we agree with Thomasson in rejecting the anti-realist and quantifier variance interpretations of Carnap's view, as well as those that depend on making the external/internal distinction rely on a verificationist theory of meaning. In Sect. 5.4, we then compare the two approaches discussed in the previous section: Thomasson's and Carnap's. We try to emphasize some subtle but significant differences between them. In the final section, we defend our own position in relation to the ontology of science, which we regard as emerging from Carnap's original ideas even more than from Thomasson's. We also highlight some particular issues related to recent interpretations of Carnap that we dismiss as misguided.

5.2 From Fiction to Scientific Discourse

How should we treat fictional discourse? What kind of thing, if any, is a fictional character? Moreover, what is fiction per se? What is the difference between fiction and non-fiction? Or, in other words, what makes discourse fictional? How should we evaluate statements *about* a work of fiction and how should we evaluate them *within* that fiction? All these are interesting questions that have attracted a lot of philosophical attention in recent decades (see Searle, 1975, van Inwagen, 1977 and 1983, Zalta, 1983 and 1988, Currie, 1990, Walton, 1990, Thomasson, 1999, Deutsch, 2000, Hofweber & Everett, 2000, Davies, 2001, Voltolini, 2006, Friend, 2007, Sainsbury, 2010, Landy, 2012, Everett, 2013, Matravers, 2014, Stock, 2017, Woods, 2018, Currie 2020, Friend, forthcoming). In fact, negative existential statements constituted a traditional problem of the philosophy of language, or the so-called puzzle of nonexistent objects.

As far as the issue of the nature of fictional characters is concerned, one option is to deny their existence: fictional names (in general, fictional expressions) do not refer at all; there are no fictional entities. But there are other options that accept fictional entities, though they differ in the way they do so. One first kind of position consists of saying that fictional characters (or, in general, individuals or classes of individuals that are only conceived in our imagination) are *possibilia* (see Lewis, 1983). A second option is to maintain, with Meinong (1904/1960), that they are nonexistent objects that nevertheless have a certain sort of being (famously, Parsons, 1980 endorsed this view in contemporary philosophy; see also Jacquette, 1996). Then there is a third way, which is committed to quantification over (or the existence of) fictional characters as abstract entities. Zalta's axiomatic theory of abstract objects (1983 and 1988), for example, allows for precisely such quantifying over fictional characters as abstract objects. Though fictional characters are, strictly speaking, nonexistent in the sense of not being spatiotemporal concrete entities (in fact, in Zalta's view, it is not possible for them to be concrete: Zalta's abstract/concrete distinction is a modal distinction), the theory quantifies over them: there is a sort of quantificational commitment to them. This is possible because, in Zalta's theory, one thing is the existential quantifier while the predicate of spatiotemporal existence is quite another (only predicable of concrete objects). Furthermore, Zalta's theory allows for a double mode of predication: exemplifying and encoding. Only abstract objects encode properties. As a general principle, for every expressible condition on properties, there is an abstract object that encodes all the properties that meet the condition. No concrete objects encode properties. However, both abstract and concrete objects exemplify properties. In this way, an abstract object is identified by the properties it encodes. In the case of fiction, fictional characters may be said to be abstract objects that encode all the properties that precisely those characters exemplify within the story.[1]

[1] In Zalta's theory, Sherlock Holmes, the fictional character native to the Conan Doyle's stories (where 'native to…' means 'that has its origin in… '), encodes, for example, the properties of being

Zalta's abstract objects theory (henceforth ZAOT) has often been associated with Platonism (see, for example, Balaguer, 2025). This connection gains some support when Zalta (1983, 52) introduces the distinction between real existence for ordinary concrete objects that exist spatiotemporally, and "Platonist existence", as a property that all and only abstract objects exemplify. According to this distinction, abstract objects are necessary: they have being in every possible world, which is a typical Platonist tenet. As Zalta (1983, 52) recognizes, there are at least two possible interpretations of ZAOT: one is really Meinongian, in the sense that we would quantify over abstract objects without committing ourselves to their existence (an expression that would be reserved for the spatiotemporal sense of "exists"); while another is the Platonist interpretation, in which Platonist existence, using the existential quantifier (∃), is explicitly affirmed of abstract objects.[2] In any case, we must also take into account that, in ZAOT, abstract objects are identified through the properties they encode. In some cases, as with fictional characters that originate in a story and entities that originate in some theory, the properties encoded by an abstract object are fixed or made explicit once the story or the theory have been formulated (or a series of speech-acts has been uttered). So, Zaltian abstract objects lack spatiotemporal existence, but in some relevant sense at least certain abstract objects are temporal (they originate with a story or with a theory) and dependent on concrete entities (they are created by people and depend on their conceptions and formulations). In this respect, there is a certain tension in ZAOT between the alleged Platonic character of abstract objects as necessary objects, and the fact that they sometimes (as in the case of fictional characters) depend on contingent entities such as authors. This is where Zalta's theory comes closest to the artifactualist account, as we are going to see.

Thomasson develops an artifactualist account of fiction according to which fictional characters, such as those that are typical in literature, are understood as abstract artifacts dependent on ordinary objects such as exemplars of texts and authors. As Thomasson herself puts it, "fictional characters are dependent objects, requiring for their very existence such entities as literary works and the creative acts of an author." (Thomasson, 1999, 24). For this reason, she speaks of the "existential

---

a detective and living in Baker Street in London, because these properties are attributed to this character in the stories, though he does not exemplify these properties. As an abstract object, Sherlock Holmes rather exemplifies the properties of being abstract and being a fictional character authored by Conan Doyle.

[2] Zalta's theory differentiates between a quantifier, ∃, and a predicate, *E*!. According to the Meinongnian interpretation, ∃ means 'there is' and *E*! means 'existence'; and we can say "there is an object that is non-existent" (i.e., $\exists x \neg E!x$: a formula that follows from the axioms of the theory). According to the second interpretation, which in some sense is more Quinean, ∃ means 'existence' and *E*! 'concreteness', and we can say, "there is/exists an object that is not concrete" (i.e., the same formula again, $\exists x \neg E!x$, but with a different meaning). The second interpretation is Platonist, since it assumes the existence of non-concrete objects; such objects are, by definition, abstract. Zalta and Linsky (1995) defend a form of Platonism called "Platonized naturalism", which they consider to be consistent with naturalism and they want to distinguish from other, more traditional forms of Platonism (even the naturalized Platonism of Quine).

dependence" of fictional characters. In the same work, she considers other *cultural and social entities*, such as artworks and entities that arise as a product of institutional and social conventions, to be examples of abstract artifacts. Social and cultural entities seem to lie between concrete material objects and mental states: they cannot be identified with either, but are dependent on collective intentionality and social practices more than on particular mental states of single isolated individuals (ibid., 128). Thomasson explicitly acknowledges the peculiar character of all these abstract objects: they are artifacts in the sense that they are dependent on our acts of creation, on our conventions and social practices, on our intentional states; and hence they are dependent on contingent entities such as humans and human groups or institutions. Thus, they are neither necessary objects (as they are in Zalta's view) nor merely possible entities; rather, they exist as long as the creative acts, and practices on which they depend take place (see ibid., 38–40).

Thomasson's approach to fiction and fictional characters should be meticulously distinguished from a well-known approach to fictional discourse that has been one of the most attractive in recent years, namely, the make-believe account introduced by Kendall Walton (1990), also called the pretense theory. This has been a popular approach among those philosophers who do not want to commit themselves to postulating the existence of fictional entities and find the classical ways of treating fictional names, inspired by Frege and Russell, clearly insufficient. The theory seems to work very well when considering statements that are made within a fiction, but it does not fare so well when we make statements about fictional characters *outside* the fiction (metafictional sentences), or when we make statements about real people that compare them to fictional characters (mixed sentences). Thomasson (1999, 97) argues that, in order to give a better account of fictional discourse, we ought to postulate and commit ourselves to fictional characters as abstract, yet artifactual, entities.[3] And this is precisely what she does in her theory.

It is quite interesting that Thomasson even mentions scientific theories and laws among the possible examples of abstract artifacts in several passages in her book (see, for example, ibid., 135, 144–146). However, it is not until Thomasson (2020) that she focuses on the case of science and proposes something explicit in relation to scientific models. She rightly points out that "artifactualist approaches to fiction have largely been overlooked by those aiming to understand scientific models on analogy with fiction", as is the case in Frigg (2010a, 2010b, 2010c), Frigg and Nguyen (2016) and Toon (2010a, 2010b, 2012). These authors have popularized Walton's make-believe approach as a way of understanding and making room for scientific fictions. Thomasson goes as far as to argue that authors who defend treating scientific models as *fictions* and favor Walton's pretense account would do better

[3] This is one of the issues that have received most attention in the recent literature on fiction, and it is not our aim to discuss it here. Nevertheless, we want to make it clear that we agree with Thomasson. In de Donato-Rodríguez and Falguera (2020, see mainly Sect. 8.5, 165–170), we already provided several arguments related to external, internal and transfictional sentences that go a long way towards showing why, in the end, we need to commit to fictional entities (as abstract artifacts).

to abandon such a view in favor of considering models to be abstract artifacts, hence embracing artifactualism with regard to models.

When we pursue this line of thought relating scientific theories to fictional discourse, to a certain extent it is usual, and perhaps also natural, to consider that, in the case of theories that have been rejected and whose theoretical terms are said to be vacuous (in the sense of not referring to an actual entity), the terms do in fact refer to fictional entities understood as abstract artifacts (instead of not referring at all). There is certainly an element of truth in saying that false (or rejected) theories are like fictional stories, in a certain sense: they do not speak of anything real, they are just positing certain entities, not of course for people's enjoyment, but in order to explain and predict phenomena, which is what theories are for.[4] What is by no means usual is to think that theoretical terms in general, even those posited in theories that are currently accepted, refer to abstract artifacts too. This is, nonetheless, precisely what we argue for in de Donato-Rodríguez and Falguera (2016 and 2020), appealing to the idealized character of scientific theories (and models), with counterfactual assumptions and approximations of intended applications. This is in no way the same as eliminating the difference between terms in rejected theories and those posited in currently accepted theories, as the difference lies in the different function and success they are assumed to have.

The idea that scientific theories and models are the product of multiple idealizations, counterfactual assumptions and approximations, is not new or strange at all. One legitimate option here is to treat them analogously to how we treat fiction and fictional entities (see Vaihinger, 1911, Duhem, 1914, van Fraassen, 1980, Cartwright, 1983, and 1999, Fine, 1993, Baberousse & Ludwig, 2009, Suárez, 2009, Woods, 2018) and—why not?—as abstract artifacts. This is a genuine option, and is in line with considerations in Thomasson (1999), since theories and models are good candidates for being artifacts: constructs made by humans in order to accomplish certain cognitive functions.[5] A further step would be to consider theoretical entities as another example. Clearly, theoretical entities would, in this sense, be considered as abstract objects that are determined in virtue of the theoretical core conditions that, according to the theory, they are to meet. In terms of ZAOT, it would be very natural to proceed this way, as we argue in de Donato-Rodríguez and Falguera (2016). Similarly to the way in which in ZAOT we have the operator "in the story", we may introduce the operator "in the theory" as a way of speaking of all the properties that are attributed in the theory to the theoretical entities that are posited (and, analogously, of everything that is true within the theory).[6] In terms of Thomasson's arti-

[4] Zalta (1983, 146) explicitly considers this when he puts forward the example of absolute simultaneity as an example of a fictional (scientific, theoretical) relation. So, ZAOT can admittedly be applied to this and other similar cases (ether, phlogiston, Vulcan, etc.).

[5] Laws could also be considered good candidates, as examples of rules that are imposed on nature by a collective decision also for their cognitively relevant virtues (explanatory, predictive, heuristic, etc.).

[6] In fact, in de Donato-Rodríguez and Falguera (2016, 24–25) we propose understanding scientific theories as a subtype of stories (which are in turn a subtype of situation, with both stories and situations being subtypes of abstract objects that can be characterized in terms of ZAOT).

factualism, it would also be possible to understand this kind of object as cultural constructs having particular cognitive functions and definite properties that are attributed to them in theoretical discourse.

What we would gain by making this move is a homogenous and coherent theoretical framework for scientific theories and models, and all their components: laws and theoretical concepts, for example. One of our arguments for concluding that it is better to treat scientific models and theories as abstract artifacts is to a certain extent similar to Thomasson's point when she argues that cultural products, and in particular novels and certain artworks, are abstract artifacts. This is so because theories are not to be confused with any of their linguistic formulations nor are scientific models to be confused with the material supports by means of which they are expressed (three-dimensional representations, diagrams, graphics, mathematical formulations, linguistic formulations, etc.), just as novels are not to be confused with any of their copies. Scientific theories and models are intensional contents. As intensional entities they are better thought of as abstract objects, and more particularly as abstract artifacts as long as they depend on human agents. We reinforce this idea by showing how supposing them to be abstract artifacts helps to account for certain intuitions and practices related to the semantics of these notions (see de Donato-Rodríguez and Falguera 2020 for details, mainly Sects. 8.3, 8.4, 8.5).

5.3 The Neo-Carnapian Approach to Ontology and Meta-Ontology

As we have seen, the notion of 'abstract artifact' is introduced by Thomasson (1999) to consider abstract objects which are created by human beings: works of fiction (novels, stories, etc.), fictional characters, institutions, laws, and so on. They are non-spatial (and, of course, non-mental), though they can be seen as temporal entities, as long as they are authored by human beings at a given moment. Quite surprisingly, Thomasson's artifactualism is not usually considered in the context of debate on the ontology of science or, more particularly, in the context of the realism/instrumentalism debate concerning the interpretation of scientific theories.[7] Similarly, Thomasson's artifactualism is not usually related to her contribution to the easy approach to ontology and meta-ontology (see Thomasson, 2015), even in her own work. So we will now consider Thomasson's view of and work on this easy approach, before going on to relate this to her work on fictions and artifactualism.

[7] Some exceptions are de Donato-Rodríguez and Falguera (2016) and (2020), and Thomasson (2020).

5.3.1 The Easy Approach

In a nutshell, what makes the easy approach "easy" is the epistemology, as Thomasson (2019, 223) herself recognizes. The core idea is that questions of existence cannot be resolved by appealing to substantive criteria of existence that may turn the issue into an ontological discovery of essences or the real nature of the world. Thomasson's approach is incompatible with hard ontology; for example, with the sort of Quinean approach to ontology that makes ontological questions "substantial". As opposed to the hard ontologist, Thomasson refuses to take ontological questions as substantial. Specifically, she rejects the idea that there is a general substantive criterion for existence, no matter whether that is grounded on causal efficacy, mental-independence or observability, or simply posited by our best scientific theories (see Thomasson, 2015, mainly Sect. 2.5, 115–121). This is so because there is no further aspect related to "there is" and "exists" more than the purely formal concept of existence, which according to her is given by the rule "Ks exist iff the application conditions actually associated with 'K' are fulfilled", which is called the core rule or schema (E) for existence (see Thomasson, 2015, 86).

What Thomasson calls "application conditions" are basically understood as semantic rules that competent speakers should master. They need not be descriptive, need not be necessary and sufficient, and may require deference to experts and to the world (Thomasson, 2015, 91).[8] Depending on the subject matter, that is, depending on the kind of question, the answer may only involve logic and conceptual knowledge, or it may also involve empirical enquiry. When the answer just involves logic or conceptual knowledge, any of us could answer the question practically automatically or without much effort. But if the existence question is empirical, the answer may not come so quickly, or may involve appeal to the relevant experts. This is because, according to Thomasson, to answer certain existence questions (from ordinary questions like: "Is there a polar bear in the zoo?" to more sophisticated ones like "Is there a photon that was produced by the impact of the two particles in the accelerator?") we need to make a series of empirical or observational inquiries. Furthermore, in certain cases we need to ask the relevant experts who allegedly know the best conditions needed for the target to be a member of the relevant kind and how it is to be determined whether the target really meets the conditions. Thus, the "easy" in the easy approach does not necessarily mean "trivial" nor "immediately or quickly established, determined, known, etc.". The easy approach is only "easy" in the sense that existential questions are understood as straightforwardly answerable by conceptual or empirical means, which are not problematic in any case, that are given by the associated conditions.

The approach implies that, at least in many cases, a common speaker, just by using linguistic and conceptual competence alone, plus perhaps some observational

[8] It is very important to add a further restriction on application conditions, namely that they are not 'circular' in the sense that they make explicit reference to the existence of the K's (see Thomasson, 2015, Sect. 2.3, 96ff.).

or empirical skills, would probably be able to answer the existence question in a straightforward and often positive way. In order to derive an ordinary ontological claim, let us say, like "there is a chair behind you", we only need to appeal to an uncontroversial claim like "there are particles arranged *chairwise*, there behind you" and to a conceptual truth provided by the application condition associated with the term "chair" that would be something like "if anywhere there are particles arranged chairwise, then there is a chair in that place". Typically, ontological debates like those referring to the existence of properties, propositions, events, numbers, or fictional characters are no exception. Thomasson thinks, and this is genuinely characteristic of her approach, that this way of understanding existence questions can be extended to all kinds of ontological discussions. And this is so, again, because of the fact that there is no further aspect related to expressions like "there is" and "exists", beyond the formal concept that is given by schema (E). So, if we want to know whether there are propositions, we only need to move from the undisputed claim that the woods are green and from the conceptual truth that "whenever P is the case, then the proposition *that P* is true" to the derived claim that the proposition *that the woods are green* is true; and finally to the ontological claim that there is a proposition, namely *that the woods are green.*[9]

What Thomasson's approach results in is a double-edged consequence of philosophical relevance. In the first place, we have what she calls "simple realism", according to which we may arrive at affirming the existence of ordinary objects and other philosophically controversial entities like numbers, events, or propositions (see Thomasson, 2015, Sect. 3.3.). In the second place, we have a meta-ontological deflationism that consists of maintaining that existence questions are always straightforwardly answerable (by conceptual or empirical means) and that nothing more is left for any sort of metaphysical epistemology that may try to investigate the supposedly "deep nature" of the world and of the entities that populate it (see Thomasson, 2015, Sect. 3.4, 158). Once we have said this, we should perhaps add that, although simple realism and meta-ontological deflationism are both consequences of her approach to the notion of existence, simple realism and meta-ontological deflationism are in fact orthogonal theses. One could maintain simple realism without accepting meta-ontological deflationism. This option has recently

[9] As Thomasson (2015, 132–145) herself recalls, it is important to take into account that her easy approach contains obvious reminiscences of both neo-Fregeanism (Hale & Wright, 2001, 2009) and Schiffer's pleonastic view (2003). But there are also important differences. On the one hand, neo-Fregeans start from an identity statement or equivalence principle as the conceptual truth that is necessary to derive the relevant claim (Thomasson does not require such equivalence principles). On the other, Thomasson's approach is, in an important sense, also more general than Schiffer's pleonastic approach, due to fulfilment of the application conditions for 'K' being guaranteed, even if the sentences whose truth we have to establish do not make explicit mention of 'K'. "[T]he trivial arguments of Schiffer (and the neo-Fregeans) for the existence of disputed entities [can be seen] as special cases of the use of the easy approach to existence questions" (Thomasson, 2015, 139–140). Another possible difference is that, according to Thomasson (ibid., 141), in her approach there is no need to use as an explicit premise a conceptually distinct statement in order to answer the existence question easily.

been explored by Korman (2019), who argues in favor of the idea that Thomasson's trivial inferences can be a legitimate route to affirm the existence of entities that are philosophically controversial, but at the same time he criticizes the way in which Thomasson defends her meta-ontological stance (his argument has to do with the fact that he considers her doctrines on the different senses of "object" to be doubtful). Meanwhile, if we think that, *prima facie*, her deflationism with respect to meta-ontology is a sufficient condition for defending simple realism, we should take into account that, from a strictly Carnapian point of view, one could be meta-ontologically deflationist without committing oneself to Thomasson's simple realism (at least understood in a certain way: that some putative entities are real in an external sense, though they are not *the substantive entities* that the traditional metaphysics or the neo-Quinean approaches try to establish).[10]

Before we pass on to the next section, let us just make an observation in relation to the easy approach and artifactualism. Mentions of fictional characters and abstract artifacts in Thomasson (2015) are mainly incidental and nothing substantive or new is said in the book about the contents or main theses of Thomasson (1999). One thing does, however, seem clear ~~to clear~~: what she says regarding existential questions in the later book can of course be applied to fictional characters, since they are abstract artifacts. That is, if Thomasson is correct in her easy approach, the existence of fictional characters (as a set of abstract artifacts) can also be easily derived from the application conditions for fictional terms. It is one thing is to reject the existence of fictional characters as real people, which is just complete commonsense, and quite another to (easily) infer the existence of fictional characters as abstract artifacts; an inference that we would indeed be authorized to make if we embrace Thomasson's view.

5.3.2 Carnap's Approach to Ontology and Meta-Ontology

The easy approach is, according to Thomasson (2015), a neo-Carnapian approach. Before we focus on a comparison this approach and the class it allegedly belongs to, let us first recall some of the highlights of Carnap's original proposal. Several authors do in fact consider themselves to be neo-Carnapian in some sense or another, and in recent times there has been a lot of discussion in meta-ontology that represents a revival of the old debate between Quine and Carnap about ontology (see: Chalmers et al., 2009; Blatti & Lapointe, 2016). The two views that clash here are, on the one hand, the Quinean "hard" view of ontology, and on the other, the Carnapian deflationary view. "Deflationary" and "Carnapian" are not synonymous, but in the literature addressing meta-ontology, they have usually been considered to

[10] There is a certain tension, or so we believe, contained within the way in which she characterizes her so-called "simple realism". From a Carnapian point of view (as we understand it), it is not justified to speak of any form of realism, and even if we understand it in a purely deflationary way, use of the word "realism" may lead to confusion. More on this in Sect. 5.4.

be almost interchangeable.[11] Carnap himself never used the term "deflationary" to describe his position regarding ontology (or any other of his positions); but Thomasson does describe his meta-ontological views as "deflationary" and links them to a deflationary approach to truth and reference (Thomasson, 2015, 87). Such a deflationary approach to the concepts of truth and reference is probably what is most coherent and natural once we adopt a deflationary approach to existence.[12] What is involved in such deflation is the idea that terms like "existence", "reference", and "truth" are not naming substantive properties or relations that we ought to discover or whose essence should be revealed by our metaphysical enquiries (see ibid., 88).[13]

What a deflationary approach to existence says is that there is nothing more, nothing substantive, to be added to the notion of existence besides the formal, second-order concept that Frege helped to define and Carnap, as a good pupil of Frege, accepted and vindicated. An analogy can be drawn here with the deflationary approach to truth, according to which there is nothing more in relation to the concept of truth than the T-schema for propositions, *<p > is true iff p* (or a disquotitional schema for sentences: *'s' is true iff s*), inspired by the Tarski biconditionals of Convention T (see Stoljar & Damnjanovic, 2014). In this sense, exactly as the easy approach to ontology rejects any characterization of existence based on substantive criteria such as causal efficacy or being posited by our best scientific theories, so the deflationary approach to truth rejects any elucidation of the concept of truth that goes beyond the T-schema, something of the kind "*x* is true iff *x* is such-and-such" or "*x* is true iff *x* has this or that property"; and this is precisely why the position is named "deflationary" (see Horwich, 1999, 10–11). The T-schema says all that is needed, all that is relevant to the notion of truth. If this is what we mean by deflationary, then Carnap was also a deflationist about existence, because, of course, he also rejected any substantive, metaphysical criteria for existence. All we know about existence is the formal notion that is captured by the existential quantifier, and is also expressed by ordinary expressions like "there is" or "exists". The meaning of the quantifiers over all the different languages (frameworks) remains the same and

[11] See, nevertheless, Hofweber (2016) and Kraut (2016), who cast doubt on the alleged "deflationism" of Carnap. Hofweber does this on the basis of a reinterpretation of the internal/external distinction which we do not agree with (which would allegedly render Carnap's views compatible with traditional metaphysics; something we believe to be inadmissible). Kraut rejects the deflationist interpretation of Carnap and tries to substitute it by an expressivist one, which is far more defendable, but still debatable.

[12] In fact, Thomasson (2014) claims that deflationism with regard to truth entails deflationism with regard to existence.

[13] Meanwhile, it is also true that there have been other 'easy' approaches to ontology which, though somehow related to Carnap, can only be considered as strictly Carnapian with difficulty; and in some cases, they are strictly speaking non-Carnapian, namely those of the neo-Fregeans (Hale & Wright, 2001 and 2009), Schiffer's pleonastic approach (1994, 1996 and 2003), and two recent approaches to the ontology of mathematics: Rayo's trivialist Platonism (Rayo, 2013) and Linnebo's minimalist approach to 'thin' objects (Linnebo, 2018). Some of these approaches are considered by Thomasson (2015, Chap. 3) and compared with her own account.

does not vary (as opposed to Hirsch's quantifier variance interpretation[14]). What changes from framework to framework is obviously the domain of the entities that are the semantic values of the variables of the language.

It is convenient now to recap, briefly, Carnap's ideas with regard to ontology. He carefully distinguished between internal and thus properly meaningful existence questions, on one hand, and external, meaningless existence questions, on the other. Internal questions are always relative to an available linguistic framework with proper rules governing the use of the terms of that language. Examples of systems for linguistic frameworks are the system of things, the system of natural numbers, the system of real numbers, the system of propositions, and the spacetime coordinate system in physics. Existence questions that go far beyond linguistic frameworks (and hence are external questions with regard to them) do not have cognitive value at all and should be considered pseudo-questions. Internal questions, however, do have cognitive value; but in return, they receive a non-problematic answer within the linguistic framework, just as in the case of Thomasson's easy inferences to existence claims. The non-problematic, uncontroversial answer to an existence question turns out to be an answer to be determined by logical or conceptual methods alone in some cases, or in some other cases, by empirical means. This will of course depend "upon whether the framework is a logical or a factual one." (Carnap, 1950b, 206).

An immediate consequence of this view is that any external claim of existence and any assertion about the real entities that allegedly constitute our external world are devoid of cognitive content. Turning back to science, this is another way of saying that scientific theories cannot help to establish what the real ontology of the world is, what the real essences of things are. Neither Platonism nor epistemic realism (indeed, any kind of realism in the sense of the hard ontologist) nor any anti-realism with regard to abstract or concrete entities, as traditionally conceived, can in the end be maintained, as the questions they try to answer are not answerable (i.e., not decidable by empirical or analytical means).[15] So, the only way to accept entities such as properties, propositions, or numbers is to accept their existence in relation (internally) to a certain linguistic framework. As Carnap famously said, to admit a

---

[14] More on this issue in Sect. 5.4 below.

[15] For this reason, any attribution of anti-realist positions to Carnap are completely misguided, as are those of the quantifier variance or verificationist interpretations. To reject Carnap's views on ontology on the grounds of avoiding his alleged anti-realism or of his alleged verificationism, cannot be legitimate. Carnap himself had already abandoned verificationism as a theory of meaning in favor of a more pragmatic and realistic position, confirmationism, on the basis of the notion of probability—and this, as early as Carnap (1936–1937) (see also Carnap, 1950a). As far as anti-realism is concerned, as it is usually understood (as the position that abstract objects, fictional characters, numbers, propositions, etc. do not exist; i.e., the ontology of the world does not contain such things, as a matter of fact), it cannot be attributed to Carnap, because that would mean entering the field of hard metaphysics. If, however, following David Chalmers, we understand "(ontological) anti-realism" as the position that there is no fact of the matter as to which framework is correct (not just useful, but correct in the cognitive, non-pragmatic, sense), then in this particular sense Carnap was an anti-realist (see Chalmers, 2009, 78).

new kind of entity is in fact to accept a certain linguistic framework: "We may still speak (and have done so) of 'the acceptance of the new entities' since this form of speech is customary; but one must keep in mind that this phrase does not mean for us anything more than acceptance of the new framework; i.e., of the new linguistic forms. Above all, it must not be interpreted as referring to an assumption, belief, or assertion of 'the reality of the entities'." (Carnap, 1950b, 214) External questions about a certain kind of entities are only acceptable when they turn out to be interpreted as questions about how useful (for given purposes) the linguistic frameworks are, that is, as *pragmatic* questions about which linguistic frameworks are more useful and for which purposes.

5.4 Comparing the Easy Approach with Carnap's Views

As we have already seen, Thomasson (2015) considers her own view to be a neo-Carnapian approach to existence questions. Though she explicitly says that her easy approach is an heir of Carnap's meta-ontological views, at the same time she insists that her account is not tied to any form of anti-realism or to positions committed to the idea of quantifier variance (favored, for example, by Hirsch, 2002, 2009, 2011; see also Putnam, 1987, and Sider, 2007). Thomasson insists on this for good reasons: according to her, the inaccurate association that has usually been made between Carnap's approach and positions such as verificationism, anti-realism and quantifier variance is the principal reason why his ideas have been dismissed. What Thomasson proposes is completely devoid of any such associations and she can still claim that her view is Carnapian. On the one hand, she points out, correctly to our mind, that for Carnap, as well as for his teacher Frege, 'exists' together with 'there is' is a univocal, formal term, having a correct and natural paraphrase in the first-order existential quantifier. (In fact, it is a second-order concept, like that of number, if Frege is right, rather than a first-order predicate.) This can be seen when we acknowledge that Carnap's way of treating ontological disputes has nothing to do with saying something along the lines of: ontological disputants are equally right in making their existential affirmations because what each of them means by "existence" are completely different things. As Thomasson (2015, 161) reminds us, there can be different approaches to ontological disputes. What she holds is that ontological disputes are easily resolved. So, strictly speaking, it is not worth entering the debate if we expect to have substantive answers as the hard ontologist wishes to find. Others maintain that we will never know the answer, a kind of "*ignoramus et ignorabimus*" (see, for example, Bennett, 2009), or that there is no fact of the matter that may help to decide the issue: ontological statements do not have any truth-value (cf. Yablo, 2009).

Neither Thomasson's nor Carnap's view should be confused with any of the forgoing positions. Thomasson would not deny that ontological questions can be answered, many times positively, albeit easily; and Carnap would admit that they have a cognitive content and can be answered, although merely internally. Even externally, they could be accepted as long as we transform them into pragmatic, non-cognitive questions about whether we use one language or another. In effect, philosophers formulate questions that the ordinary person or the scientist never raises: they seriously ask whether numbers or propositions really exist, without acknowledging that "the controversy goes on for centuries" and cannot be solved (Carnap, 1950b, 207). When, later in the same text, Carnap considers the debate between the Platonist and the nominalist in the philosophy of mathematics, he says: "I cannot think of any possible evidence that would be regarded as relevant by both philosophers, and therefore, if actually found, would decide the controversy or at least make one of the opposite theses more probable than the other." (Carnap, 1950b, 219) This is similar to what Yablo, for example, has in mind; but it is not identical, because Carnap is saying this only of existential questions as considered externally. Even when considered externally, they can still be maintained just as questions about the convenience or otherwise of using one framework that contains certain terms instead of another one that avoids them. In the end, a neopositivist like Carnap could accept abstract entities, towards which empiricists always had qualms, just in terms of "the ratio of the results achieved to the amount and complexity of the efforts required." (Carnap, 1950b, 221).

Interestingly enough, Hirsch's recommendation (2011, Chap. 9) for resolving a verbal dispute cannot be followed in most of the interesting and controversial cases, like the Platonism versus nominalism debate in the philosophy of mathematics, to continue with the same example. Hirsch's protocol of searching for a way to paraphrase one's opponent's claim in terms of the vocabulary that one uses and that may be accepted as true, is not always a possibility, and it does not work in this particular case.[16] In Carnapian terms, the only way of settling the issue, when the dispute is between opposite frameworks trying to resolve the same problem, would be to see whether adopting the opposing stance makes any difference or provides any advantages in terms of instrumental efficiency. There is no fact of the matter that can help to settle the issue when it is viewed at the external level (in its non-pragmatic consideration). So the dispute is not exactly verbal, if by that we mean that the problem lies merely in the words (see Chalmers, 2011 for a plausible analysis of the concept of a merely verbal dispute). The problem, in Carnapian terms, is that when considered externally, we cannot find a resolution of the debate; rather, we have to settle for rethinking the issue on a pragmatic level, which is certainly no trifle either.

Even if having they bear similarities, there are important differences between the approaches of Thomasson and of Carnap. The first thing to point out is that, contrary to Carnap's approach, Thomasson's is concerned with existence questions that are asked *in ordinary language*, as put by ordinary, yet competent, speakers (although sometimes the criteria of experts are required): Carnap considers existence questions always in relation to linguistic frameworks of some kind, guided by the tolerance principle (with the constraint of cognitive significance). This is not overlooked by Thomasson, who acknowledges that Carnap is primary interested in formal,

[16] Hirsch (2011, 162) himself realizes this.

technical languages, whereas she is concerned with ontological questions as formulated in ordinary language (Thomasson, 2015, 44). Even when Carnap explicitly considers the so-called 'thing-language' as an example of his linguistic frameworks, this is an idealized and non-privileged version of ordinary language, restricted to observational things and events (that is, he accepts the use of it as a specific framework to settle issues that are about ordinary macroscopic objects, but this is just one framework among others). For Carnap, ontological questions become relevant when we put them in a principled way, in the correct formulation of a linguistic framework. It is true, according to Carnap, that we can discuss the existence of an entity in the ordinary sense of the thing-language, but in order to discuss other, non-ordinary examples, we must turn to the relevant framework. This is far more appropriate when we explicitly consider examples of scientific discourse, such as the terms in physical or biological theories. The same goes for terms like "number", "proposition" or "property": we need some specific linguistic framework which introduces them beyond ordinary language. To rely merely on the competence and abilities of normal speakers, on their linguistic skills and conceptual intuitions, as Thomasson expects, is perhaps not the best way to treat debates that have attracted the attention of philosophers for many centuries. As Button (2020, Sects. 2.2. and 3.4.) has correctly pointed out, ordinary language is a complicated phenomenon and arguably can only be tackled by invoking a context principle, which probably means that the task of answering the relevant questions will not be so easy. Perhaps in some cases even intuitions will point in different directions simultaneously; something that can easily be confirmed when we see that opposite metaphysical views can always find some support from our conceptual and linguistic intuitions, even though recent contributions to experimental metaphysics (see Rose, 2018) may encourage more research on this. On the other hand, it does indeed sound very strange to have confidence in nothing more than trivial inferences and relevant linguistic competence in order to settle ontological disputes. Perhaps easy ontology is too easy and no philosopher who cares about ontology would entertain the feeling that, by looking at the issue through Thomasson's glasses, we can properly enter into it. Carnap instead seems to pave the way for a reconsideration of ontological disputes beyond their mere internal consideration, by encouraging us to reformulate the questions as pragmatic matters.

A further point concerns the possibility of applying Thomasson's easy approach to the case of science: a possibility we want to examine explicitly. One difficulty is the fact that theoretical terms are not introduced in contexts of ordinary language or by social conventions, as many of the terms Thomasson considers are. When she explicitly considers the case of scientific terms in her 2015 book, she seems to assume (in one of the possible interpretations) that theoretical entities are candidate concrete entities. Surprisingly, she does not seem to be aware of the possibility, much more justifiable in Carnapian terms,[17] of considering theoretical entities as a kind of abstract artifacts.[18]

Here, there arises a legitimate question that we want to ask: Are Thomasson's points of view in fact strictly Carnapian, particularly when we take a look at the scientific case? Is Thomasson assuming that the empirical sciences are providing, thanks to their best cognitive products, real entities in the sense of hard metaphysics? The answer to this question is not totally clear to us. In one sense, it seems that no: given her vindication of Carnap's considerations regarding ontology (and meta-ontology), and her rejection of hard ontology. However, in another sense, it may seem that she is. Thomasson sometimes speaks as if easy ontology is constrained to the postulated entities of ordinary speech, whereas her discourse on entities considered by the empirical sciences seems to be along (hard) realistic lines. Thomasson seems to have something like this in mind, given the way she characterizes the notion of the "application conditions" of a general noun *K*. In explaining the feature which means that application conditions may involve deference to experts or to the world, she argues:

> This thesis is intended to make it clear that one can accept application conditions while also accepting the validity of the main arguments that have led to the popularity of externalist and causal theories of reference for natural kind terms. [...] Where terms have world-deferential rules like these, one may then defer to the world (or to experts who have specialized knowledge of the world) to discover exactly what it is that distinguishes the kind *platypus* from other biological kinds, what its essential features are, and so on. (Thomasson, 2015, 94)

And later in the text, she clarifies that, for observational terms, the application of a term may be justified by the presence of certain visible conditions, while for natural kind terms, it may require the presence of entities of the same kind (ibid., 96, n.13). Where Thomasson speaks of "natural kind terms", it would perhaps be better to speak of "theoretical terms", which is far more neutral from an ontological point of view. When we speak of "theoretical terms", it is still *prima facie* and strictly speaking an open question whether they refer to concrete or to abstract objects, or even to fictions ones. We could even say, taking up van Fraassen's (1980) position, that we do not need to settle the issue, because we do not need to commit ourselves to the existence of theoretical entities (understood as concrete entities, as is usual) when we accept a scientific theory. But, in speaking of "natural kind terms" and in speaking of them as she does, Thomasson seems to enter the field of hard ontology. Her appeal to the causal theory of reference is another clue that she is straying into the

[17] See, above all, Carnap (1956).

[18] As we will see, Carnap's (1956) proposal is that the referents of theoretical terms are mathematical entities, and therefore abstract entities (artifacts, perhaps). Our view is different, though we also consider the referents of theoretical terms to be abstract entities (artifacts): not because of their mathematical nature (numbers, and sets and *n*-tuples of numbers), but rather because they are like fictional characters.

realm of hard metaphysics. In the end, her simple realism seems to go far beyond what is allowed in Carnapian terms.[19]

5.5 A Deflationary and Artifactualist Ontology for Science

We can ask whether theoretical terms in our best theories refer to real (kinds of) entities, and whether observational terms refer to real (kinds of) entities. We can also ask which kind of entities the direct (putative) referents of theoretical terms are: concrete or abstract (artifactual) objects. We have seen that Thomasson seems to think of theoretical entities as concrete entities (when they exist), but it would be more coherent in Carnapian terms to think of them as abstract objects (artifacts). We maintain that this is precisely the case, but not for the reasons Carnap had in mind.

Our proposal involves considering the linguistic formulations of scientific theories as linguistic frameworks (in the Carnapian sense). We aim to establish what kind of ontological commitment we should have once we decide, based on the evidence, to accept a scientific theory. We contend that ontological commitments related to the scientific formulation of a well-confirmed scientific theory are always internal (not external). Moreover, any other position with regard to this issue in strictly Carnapian terms is excluded. More particularly, a well-confirmed scientific theory commonly involves a commitment to the (internal) existence of three kind of entities: (i) entities designated by mathematical terms that are needed in many of our current scientific theories, (ii) entities that are designated by the observational terms of the theory, and (iii) entities that are designated by the theoretical terms of the theory. Mathematical terms are paradigmatically taken to be abstract entities; although, given our Carnapian approach, not in a Platonic sense, but in the trivial way that is internally followed by acceptance of mathematical language—and the corresponding general mathematical terms—in the formulations of our scientific theories.

Observational terms would typically be taken to refer to concrete entities—individuals, properties, or relations that are observable—though our commitment to them cannot be equally external (if we accept Carnap's stance). There is a further epistemological reason for this: they arise from how the world is phenomenologically presented to us, and not from how the world really is. Recall that what Carnap calls "the world of things" is just "the spatio-temporally ordered system of observable things and events" (Carnap, 1950b, 207), and the so-called thing-language is just one non-privileged linguistic framework among others.

Theoretical entities, that is, the referents of theoretical terms, are also to be understood, we propose, as abstract objects (in the artifactualist sense) since they

[19] Recently, Marschall (2019) also pointed out that there could be a certain tension (perhaps not insurmountable) within her approach by trying to show that her views rely on a conception of the hierarchical order of sortals that calls into question the alleged compatibility between easy ontology and (simple) realism.

are the product of idealizations and counterfactual assumptions. For us, every theoretical term, no matter whether it belongs to a theory that has been rejected or to a well-confirmed and currently accepted theory, is taken to refer to an abstract artifact. In both cases, they are well-defined objects that result from the description of a theory (we accept the descriptive approach to theoretical terms, and reject the causal theory of reference, and in particular as it is applied to theoretical terms[20]). For Carnap, theoretical entities were also abstract entities, though not for the same reason; rather, because of their mathematical nature. Carnap (1956, § IV) insists on this idea by indicating that the entities of the theoretical vocabulary are based on a system of spacetime coordinates, where spacetime points are ordered quadruples of real numbers, spacetime regions are classes of points, and physical magnitudes are functions from points or spacetime regions to real numbers or tuples of real numbers. Carnap (1958, 1966 and 2000) uses the Ramsey-sentence to provide the empirical content of a theory in a way that avoids the use of theoretical terms, but does not intend to avoid commitment to their referents, though understanding these as logico-mathematical entities.

The idea that theoretical entities are abstract objects because of their fictional nature is worth pursuing, and this may well seem a more plausible reason for those who think that Carnap's justification implies a kind of ontological reduction to merely mathematical entities that is a little strange. But if we think that scientific theories are constitutively based on highly idealized lawlike conditions and models, then theoretical expressions cannot primarily refer to concrete objects.[21] (There are further arguments, of a semantic nature, that also give us a rationale for considering theoretical entities, models and their components as abstract artifacts). In any case, they cannot refer either to external concrete objects, because these would be related to the kind of external ontological questions that are not clarified with the help of scientific theories, or to internal concrete objects, because it is not clear in a neo-Carnapian approach what these could be when they are not observational entities or concrete in the sense of how the world is phenomenologically presented to us. To speak of theoretical entities as concrete only seems to make sense in the external sense. As far as Thomasson's easy approach is concerned, it is by no means obvious what her position would be regarding the case of theoretical entities. If her position consists of considering them to be concrete entities, then this would contrast with

[20] In de Donato-Rodríguez and Falguera (2016, Sect. 2.1) we explain the reasons why causal theories of reference, even when combined with a degree of descriptivism (as in the hybrid theories of Kitcher, 1993 and Psillos, 1999), do not work in the case of theoretical terms. We also present an approach to theories, models and theoretical entities that identifies them with abstract objects in the manner of ZAOT.

[21] In de Donato-Rodríguez and Falguera (2020) we support Cartwright's dictum that models are works of fiction (see Cartwright 1983), but claiming this does not mean that scientific models are to be treated as fictional entities. We offer different reasons why it is clearly better to treat scientific models as abstract artifacts. See also Giere (2009) for a defense of the view that models are abstract objects, though we ultimately disagree with some of his theses on the relation between fiction and scientific models.

Carnap's as well as with our own view; and it could even mark an unnoticed foray into the realm of hard ontology or traditional metaphysics.

The connection between all this and Thomasson's meta-ontological deflationism now becomes clear. We align ourselves with a Thomasson-inspired position regarding ontology. Along the lines of a neo-Carnapian way of understanding ontology, we contend that ontological commitments related to our best scientific theories cannot be external. Quantification in the context of a well-confirmed scientific theory must be understood as an internal assertion regarding the existence of certain entities that are theoretically constituted and, because of this, must be considered as abstract artifacts. The entities in question would be abstract, but also as "artificial" as the cultural and social entities considered by Thomasson. And, again, no metaphysical assertion concerning the existence of these entities, yet alone concerning the probability or plausibility of their existence, as traditionally considered by hard ontologists, is admitted in this approach. The commitment to abstract entities that we are defending does not commit us to Platonism in a double sense: again, our view does not amount to an external claim regarding the existence of abstract objects. Furthermore, when we accept fictional (and mathematical) entities as abstract objects, we do not need to understand them as eternal and mind-independent objects (Platonic entities). We thus make use here of Thomasson's (1999) artifactualism, in order to understand them as a particular case of an "abstract artifact": a kind of abstract entity that depends on particular concrete entities, such as the scientists that formulate the scientific theories in which the terms for these entities occur. Furthermore, this deflationism regarding existence questions should also be applied to the notions of reference and truth; and this is, in our opinion, the option that is most coherent with Carnap's final ideas. The referents of the terms introduced by a given framework cannot be understood as the real entities the metaphysician searches for. A deflationist conception of truth arises as the most plausible, too. Of course, we are fully aware of extending Thomasson's position to the case of scientific theories.

Recently, there has been much discussion of Carnap's views on meta-ontology, to the extent that some philosophers have tried to invent their "own Carnap". Some authors, like Hofweber (2016) or Kraut (2016), question whether Carnap can be called a "deflationist", and propose reinterpretations of the external/internal distinction that are either not really Carnapian (in the case of Hofweber's idea of making the quantifier polysemic) or at best whose Carnapian credentials are debatable (Kraut's expressivist interpretation). Eklund (2009) has proposed another interpretation, that of "ontological pluralism", against which he himself goes on to present different arguments (none of them really demonstrative) and then offers an alternative or "different route". There are authors (see Alspector-Kelly, 2001) who have even blurred the alleged difference between Carnap and Quine, saying—incorrectly to our mind—that the differences are minimal and perhaps merely terminological. Against this background, we want to conclude our contribution with some remarks on what we consider to be the true character and novelty of Carnap's original ideas on meta-ontology, and to which we believe that we remain even closer than Thomasson does with her easy approach.

Contrary to recent approaches that have aimed at a naturalization of metaphysics,[22] pursuing the idea that the ontology of science and the ontology of metaphysics can come together and offer us (perhaps just "at the end of the day") real essences or the true ontology of the world, Carnap's lesson is that science and metaphysics cannot help to establish the true ontology of the world. Searching for that is an endeavor that we would do better in abandon. And the reason does not have to do with any alleged verificationism or with the kind of anti-realism that is commonly, but wrongly, attributed to Carnap. As Bradley (2018) has recently argued, Carnap's final position against metaphysics does not have to do with a verificationist theory of meaning (which, incidentally, Carnap had already abandoned),[23] but with the epistemological point that the claims of metaphysics (as traditionally understood) are not justifiable.[24] We agree with Eklund (2009) and Hofweber (2016) when they affirm that the external/internal distinction can, in principle at least, be separated from the deflationary approach; but we do not agree with the conclusions they try to extract from this. What Hofweber proposes implies an important departure from Carnap's non-cognitivist consideration of metaphysics; and Eklund's ontological maximality does not do justice to it. The latter's proposal to interpret Carnap as an ontological pluralist is quite dubious, as Carnap would not go so far as to affirm the real existence of anything we quantify over in a given framework. So, as there are many frameworks (all that the tolerance principle allows us), we would have to accept the existence of multiple kinds of things, almost as if we could "see" what exists from a privileged view over and above the frameworks. Carnap was indeed a linguistic pluralist, but not an *ontological* pluralist. Eklund's move brings him dangerously close to Putnam's and Hirsh's imaginings.[25] So, we completely agree with Thomasson (2015) when she tries to separate Carnap's approach from any verificationist, anti-realist, or quantifier variance interpretations; and we think her criticism can be extended to Eklund's view in the same way.

Finally, there has been recent discussion of how we should relate Carnap's ideas on meta-ontology to his position with regard to the realism/instrumentalism debate in the philosophy of science (see Psillos, 1999, Demopoulus, 2011, Friedman, 2011, Uebel, 2011, Lavers, 2016, Bueno, 2016, Bueno & Meier, 2019 and forthcoming). Also on this issue there have been a lot of misunderstandings. For one thing, it is not true, *pace* Lavers, that there is a tension between Carnap's external/internal distinction and his position in the realism/instrumentalism debate. His stance was that we

[22] See, for example, Ross et al. (eds., 2013) and Slater and Yudell (eds., 2017), both containing contributions along the lines of this "hard metaphysics".

[23] We already discuss this in our note 15 above.

[24] Of course, the case is different when we refer to the early Carnap from the times of the Vienna Circle, but here we are speaking of the later Carnap (1950b, 1956).

[25] In truth, Eklund's proposal is cautious: "I am inclined to embrace the former alternative [the ontological pluralism interpretation]", instead of the idea that Carnap was not, after all, an ontological pluralist (Eklund, 2009, 142). The only alternative for him is to make Carnap into a Platonist, though of *a special kind*. Unfortunately, Eklund does not continue to work along these lines, and what he ultimately proposes is a disappointing exegesis of the real Carnap.

should remain agnostic: a third alternative in that debate. Though realism and instrumentalism may seem to be two legitimate ways of talking about scientific theories, strictly speaking we cannot affirm either of them. So, Carnap's position regarding this issue was to remain agnostic.[26] This agnosticism is completely coherent with his position with respect to the Platonism/nominalism debate. In the end, according to Carnap, we have no reasons to affirm or to deny any external existence claim. To remain agnostic, or neutral (if you prefer), is to suspend judgment on this kind of claims, for no reason other than that they cannot be decided by cognitive means. Bueno and Meier (forthcoming) also perceive a certain tension between Carnap's alleged neutralism and his dismissal of metaphysics.[27] They argue that "his rejection of metaphysics as being meaningless is interestingly non-neutral. Carnap takes a particular anti-metaphysical stance on the status of metaphysics, and this violates, in part, his own neutralism." (ibid., 23) We find this affirmation not to be completely justified. It is true that Carnap dismisses metaphysical claims as non-cognitive; only in this sense could his position be labelled as non-neutral (surely it is impossible to remain neutral about absolutely everything!). But this is just in relation to external ontological claims, and there is always the possibility of reinterpreting them as pragmatic questions. Meanwhile, the "neutral neutralism" Bueno and Meier propose as an alternative is in the end very similar to Carnap's own neutralism. It involves, among other things, a neutral stance regarding frameworks and regarding external existence questions, just as the position Carnap does. The only difference seems to be the dismissal of metaphysics; but once we clarify in what sense Carnap dismisses metaphysics, we can see that the difference is not that great.

5.6 Conclusion

We have seen that, leaving some difficulties aside, Thomasson's artifactualism combined with her deflationary approach to meta-ontology may altogether serve as a useful and adequate framework to understand the ontology of science. We have also seen the extent to which our own account could count as even more Carnapian. Our

[26] This "agnosticism", as we call it, has been labeled as "neutralism" by others (see, for example, Bueno and Meier (forthcoming)). However, such labelling must be carefully understood, since other philosophy of science authors, such as Psillos (1999), Demopoulus (2011), and Friedman (2011), use "neutralism" in a slightly, but importantly, different sense when applied to Carnap's point of view in the debate between realism and instrumentalism. They use "neutralism" in the sense of the "irenic" position, as Psillos (1999, 40) calls it, according to which both realism and instrumentalism about scientific theories could be equally acceptable and compatible. Carnap's real position should be carefully distinguished from this latter view, since his position is an alternative one which suspends judgment, and is equally distanced from the previous two.

[27] Bueno and Meier (forthcoming) say that Carnap believed that metaphysics is non-rational and non-conceptual; but this is perhaps too strong. Carnap never used these terms. Strictly speaking, what he said is that metaphysics is non-cognitive, that metaphysical questions—in the form of external existence questions—cannot be answered by cognitive means (logical or empirical).

proposal of considering theoretical entities (not only models or even theories themselves) as abstract artifacts ultimately represents a novel approach in the arena of philosophical discussion that deserves to be explored.

Acknowledgments A previous version of this text was presented at the 28th edition of the Inter-University Workshop on Philosophy and Cognitive Science, focused on the work of Amie L. Thomasson, which took place at the University of Granada in December 3-5, 2019. We thank the audience and Amie Thomasson for their comments. We also thank Toffa Evans for his revision of the text and Mirco Sambrotta for his work as editor of this book. The work has been financed by the Spanish National Research Agency Projects FFI2017-82534-P and PID2020-115482GB-I00, and also by the project PICT 2018-3454 of the National Scientific and Technological Research Council (Argentina). ORCID ID of Xavier de Donato: orcid.org/0000- 0002-8464-9960; ORCID ID of J.L. Falguera: orcid.org/0000- 0003-2631-1563.".

References

Alspector-Kelly, M. (2001). On Quine on Carnap on ontology. *Philosophical Studies, 102(1), 93–122.*

Baberousse, A., & Ludwig, P. (2009). Models as fictions. In M. Suárez (Ed.), *Fictions in science. Philosophical essays in modeling and idealizations* (pp. 56–73). London/New York: Routledge.

Balaguer, M. (2025). Platonism in Metaphysics. In Edward N. Zalta & Uri Nodelman (eds.): *The Stanford Encyclopedia of Philosophy* (Spring 2025 Edition), https://plato.stanford.edu/archives/spr2025/entries/platonism/.

Bennett, K. (2009). Composition, Colocation, and Metaontology. In David Chalmers, David Manley & Ryan Wasserman (Eds.), *Metametaphysics: New Essays on the Foundations of Ontology* (pp. 38–76). Oxford: Oxford University Press.

Blatti, S., & Lapointe, S. (Eds.). (2016). *Ontology after Carnap*. Oxford: Oxford University Press.

Bradley, D. (2018). Carnap's epistemological critique of metaphysics. *Synthese, 195*(5), 2247–2265.

Bueno, O. (2016). Carnap, logicism, and ontological commitment. In S. Costreie (Ed.), *Early analytic philosophy—New perspectives on the tradition, the Western Ontario series of philosophy of science* (Vol. 80, pp. 337–352). Cham: Springer.

Bueno, O., & Meier, T. (2019). Structuralism, empiricism, and Newman's objection. *Principia, 23*(1), 53–67.

Bueno, O., & Meier, T. (forthcoming). Is neutralism possible? Carnap and philosophical debates.

Button, T. (2020). Deflationary metaphysics and ordinary language. *Synthese, 197*(1), 33–57.

Carnap, R. (1936–1937). Testability and meaning. *Philosophy of Science*, *3*(1936), 419–471; and *4*(1937), 1–40.

Carnap, R. (1950a). *Logical foundations of probability*. Chicago: Chicago University Press.

Carnap, R. (1950b). Empiricism, semantics and ontology. *Revue Internationale de Philosophie, 4*, 20–40. (Used English version: Carnap, R. (1956). *Meaning and necessity*. Chicago: The University of Chicago Press, pp. 205–221).

Carnap, R. (1956). The Methodological Character of Theoretical Concepts. In H. Feigl & N. Scriven (Eds.), *Minnesota studies in the philosophy of science* (Vol. I, pp. 38–76). Minneapolis: University of Minnesota Press.

Carnap, R. (1958). Beobachtungssprache und theoretische Sprache. *Dialectica, 12*, 236–248. (Vers. util.: Observational language and theoretical language. In J. Hintikka (1975). *Rudolf Carnap, logical empirist. Materials and perspectives*. Dordrecht: Reidel, pp. 75–85).

Carnap, R. (1966). *Philosophical foundations of physics*. New York/London: Basic Books.

Carnap, R. (2000). Theoretical concepts in science. *Studies in History and Philosophy of Science, 31*, 159–172.
Cartwright, N. (1983). *How the laws of physics lie*. Oxford: Oxford University Press.
Cartwright, N. (1999). *The dappled world. A study of the boundaries of science*. Cambridge: Cambridge University Press.
Chalmers, D. J. (2009). Ontological antirealism. In D. J. Chalmers, D. Manley, & R. Wasserman (Eds.), *Metametaphysics* (pp. 77–129). Oxford: Oxford University Press.
Chalmers, D. J. (2011). Verbal disputes. *Philosophical Review, 120*(4), 515–566.
Chalmers, D. J., Manley, D., & Wasserman, R. (Eds.). (2009). *Metametaphysics*. Oxford: Oxford University Press.
Currie, G. (1990). *The nature of fiction*. Cambridge: Cambridge University Press.
Currie, G. (2020). *Imagining and knowing: The shape of fiction*. Oxford: Oxford University Press.
Davies, D. (2001). Fiction. In B. N. Gaut & D. Lopes (Eds.), *The Routledge companion to aesthetics* (pp. 263–274). London: Routledge.
de Donato-Rodríguez, X. and J. L. Falguera (2016). "On Fictions, Theoretical Entities, and Ideal Objects : Applying Zalta's Abstract Objects Theory to Scientific Theories." In G. Borbone and K. Brzechczyn (Eds.), Idealization XIV: Models in Science (pp. 13-42). Brill/Rodopi.
de Donato-Rodríguez, X. and J. L. Falguera (2020). "The Nature of Scientific Models: Abstract Artifacts that Determine Fictional Systems". In J.L. Falguera and C. Martínez-Vidal (Eds.), Abstract Objects: For and Against (pp. 151–171). Springer.
Demopoulus, W. (2011). On extending 'empiricism, semantics, and ontology' to the realism/instrumentalism controversy. *The Journal of Philosophy CVIII, 12*, 647–669.
Deutsch, H. (2000). Making up stories. In T. Hofweber & A. Everett (Eds.), *Empty names, fiction, and the puzzles of nonexistence* (pp. 17–36). Stanford: CSLI Publications.
Duhem, P. (1914). *La théorie physique son objet et sa structure* (2nd ed.). Paris: Chevalier et Rivière.
Eklund, M. (2009). Carnap and ontological pluralism. In D. Chalmers, D. Manley, & R. Wasserman (Eds.), *Metametaphysics*. Oxford: Oxford University Press.
Everett, A. (2013). *The nonexistent*. Oxford: Oxford University Press.
Fine, A. (1993). Fictionalism. *Midwest Studies In Philosophy, 18*(1), 1–18.
Friedman, M. (2011). Carnap on theoretical terms: Structuralism without metaphysics. *Synthese, 180*, 249–263.
Friend, S. (2007). Fictional characters. *Philosophy Compass, 2*(2), 141–156.
Friend, S. (forthcoming). *Matters of facts and fiction*. Oxford: Oxford University Press.
Frigg, R. (2010a). Models and fiction. *Synthese, 172*(2), 251–268.
Frigg, R. (2010b). Fiction in science. In J. Woods (Ed.), *Fictions and models: New essays* (pp. 247–287). Munich: Philosophia Verlag.
Frigg, R. (2010c). Fiction and scientific representation. In R. Frigg & M. Hunter (Eds.), *Beyond mimesis and nominalism: Representation in art and science* (pp. 97–138). Dordrecht: Springer.
Frigg, R., & Nguyen, J. (2016). The fiction view of models reloaded. *The Monist, 99*, 225–242.
Giere, R. (2009). Why scientific models should not be regarded as works of fiction. In Mauricio Suárez (Ed.), *Fictions in Science: Philosophical Essays on Modeling and Idealization* (pp. 248–258). London: Routledge.
Hale, B., & Wright, C. (2001). *The reason's proper study. Essays towards a Neo-Fregean philosophy of mathematics*. Oxford: Oxford University Press.
Hale, B., & Wright, C. (2009). The metaontology of abstraction. In D. J. Chalmers, R. Wasserman, & D. Manley (Eds.), *Metametaphysics: New essays on the foundations of ontology* (pp. 178–212). Oxford: Oxford University Press.
Hirsch, E. (2002). Quantifier variance and realism. *Philosophical Issues, 12*(1), 51–73.
Hirsch, E. (2009). Carnap and the ontological pluralism. In D. M. Chalmers & R. Wasserman (Eds.), *Metametaphysics: New essays on the foundations of ontology* (pp. 130–156). Oxford: Oxford University Press.

Hirsch, E. (2011). *Quantifier variance and realism: Essays in metaontology*. Oxford: Oxford University Press.
Hofweber, T. (2016). Carnap's big idea. In S. Blatti & S. Lapointe (Eds.), *Ontology after Carnap* (pp. 13–30). Oxford: Oxford University Press.
Hofweber, T., & Everett, A. (Eds.). (2000). *Empty names, fiction, and the puzzles of non-existence* (pp. 117–148). Stanford: CSLI Publications.
Horwich, P. (1999). *Truth*. Oxford: Clarendon Press.
Jacquette, D. (1996). *Meinongian logic: The semantics of existence and nonexistence*. Berlin/New York: De Gruyter.
Kitcher, P. (1993). *The advancement of science: Science without legend, objectivity without illusions*. Oxford: Oxford University Press.
Korman, D. Z. (2019). Easy ontology without deflationary metaontology. *Philosophy and Phenomenological Research, 99*(1), 236–243.
Kraut, R. (2016). Three Carnaps on ontology. In S. Blatti & S. Lapointe (Eds.), *Ontology after Carnap*. Oxford: Oxford University Press.
Landy, J. (2012). *How to do things with fictions*. New York: Oxford University Press.
Lavers, G. (2016). Carnap on abstract and theoretical entities. In S. Blatti & S. Lapointe (Eds.), *Ontology after Carnap* (pp. 200–219). Oxford: Oxford University Press.
Lewis, D. (1983). Truth in fiction. In *Philosophical papers* (Vol. 1, pp. 261–280). Oxford: Oxford University Press.
Linnebo, Ø. (2018). *Thin Objects: An Abstractionist Account*. Oxford: Oxford University Press.
Marschall, B. (2019). Easy ontology, quantification, and realism. *Synthese* (published online: 18 Nov. 2019). https://doi.org/10.1007/s11229-019-02463-8
Matravers, D. (2014). *Fiction and narrative*. Oxford: Oxford University Press.
Meinong, A. (1904/1960). The theory of objects. In R. M. Chisholm (Ed.), *Realism and the background of phenomenology* (pp. 76–117). Glencoe, Ill.: The Free Press.
Parsons, T. (1980). *Nonexistent objects*. Westford, MA: Yale University Press.
Psillos, S. (1999). *Scientific realism. How science tracks truth*. London: Routledge.
Putnam, H. (1987). *The many faces of realism*. La Salle, Ill.: Open Court.
Rayo, A. (2013). The Construction of Logical Space. Oxford: Oxford University Press.
Rose, D. (Ed.). (2018). *Experimental metaphysics*. New York: Bloomsbury.
Ross, D., Ladyman, J., & Kincaid, H. (Eds.). (2013). *Scientific metaphysics*. Oxford: Oxford University Press.
Sainsbury, R. M. (2010). *Fiction and fictionalism*. London: Routledge.
Searle, J. (1975). The logical status of fictional discourse. *New Literary History, 6*(2), 319–332.
Schiffer, S. (1994). A Paradox of Meaning. Nous 28: 279–324.
Schiffer, S. (1996). Language-Created Language-Independent Entities. *Philosophical Topics, 24* (1) 149–67.
Schiffer, S. (2003). *The Things We Mean*. Oxford: Oxford University Press.
Sider, T. (2007). NeoFregeanism and quantifier variance. *Aristotelian Society, Supplementary, 81*, 201–232.
Slater, M. H., & Yudell, Z. (Eds.). (2017). *Metaphysics and the philosophy of science: New essays*. Oxford: Oxford University Press.
Stock, K. (2017). *Only imagine: Fiction, interpretation and imagination*. Oxford: Oxford University Press.
Stoljar, D., & Damnjanovic, N. (2014). The deflationary theory of truth. In Edward N. Zalta (Ed.), *The Stanford encyclopedia of philosophy* (Fall 2014 edition). https://plato.stanford.edu/archives/fall2014/entries/truth-deflationary/
Suárez, M. (Ed.). (2009). *Fictions in science. Philosophical essays on modelling and idealisation*. London: Routledge.
Thomasson, A. (1999). *Fiction and metaphysics*. Cambridge: Cambridge University Press.

Thomasson, A. (2014). Deflationism in semantics and metaphysics. In A. Burgess & B. Sherman (Eds.), *Metasemantics: New essays on the foundations of meaning* (pp. 185–213). Oxford: Oxford University Press.

Thomasson, A. (2015). *Ontology made easy*. Oxford: Oxford University Press.

Thomasson, A. (2019). Précis of ontology made easy. *Philosophy and Phenomenological Research, XCIX, 1*, 223–228.

Thomasson, A. L. (2020). If models were fictions, then what they would they be? In A. Levy & P. Godfrey-Smith (Eds.), *The scientific imagination*. New York: Oxford University Press.

Toon, A. (2010a). Models as make-believe. In R. Frigg & M. C. Hunter (Eds.), *Beyond mimesis and convention, Boston studies in the philosophy of science* (Vol. 262, pp. 71–96). Dordrecht: Springer.

Toon, A. (2010b). The ontology of theoretical modelling: Models as make-believe. *Synthese, 172*, 301–331.

Toon, A. (2012). *Models as make-believe. Imagination, fiction and scientific representation*. Chippenham and Eastbourne: Palgrave/Macmillan.

Uebel, T. (2011). Carnap's ramseyfications defended. *European Journal for Philosophy of Science, 1*, 71–87.

Vaihinger, H. (1911). *Die Philosophie des Als Ob*. Berlin: von Reuther & Reichard.

van Fraassen, B. C. (1980). *The scientific image*. Oxford: Oxford University Press.

van Inwagen, P. (1977). Creatures of fiction. *American Philosophical Quarterly, 14*(4), 299–308.

van Inwagen, P. (1983). Fiction and metaphysics. *Philosophy and Literature, 7*(1), 67–77.

Voltolini, A. (2006). *How ficta follow fiction*. Dordrecht: Springer.

Walton, K. L. (1990). *Mimesis as make-believe: On the foundations of the representational arts*. Cambridge and London: Harvard University Press.

Woods, J. (2018). *Truth in fiction: Rethinking its logic*. Cham: Springer.

Yablo, S. (2009). Must existence-questions have answers? In D. Chalmers et al. (Eds.), *Metametaphysics* (pp. 507–525). Oxford: Oxford University Press.

Zalta, E. (1983). *Abstract objects: An introduction to axiomatic metaphysics*. Dordrecht: Reidel.

Zalta, E. (1988). *Intensional logic and the metaphysics of intentionality*. Cambridge, MA: The MIT Press.

Zalta, E., and Linsky, B. (1995) Naturalized Platonism vs. Platonized Naturalism, *Journal of Philosophy, 92*, 525–555.

[illegible]

Chapter 6
Fictionalism and the Easy Approach to the Existence of Facts

Pablo Rychter

Abstract This paper defends the plausibility of antirealism about facts, and of fictionalism about 'fact talk'. That is, it defends the view that even when a particular object *o* exemplifies a property *P*, there is no further entity to be identified with the *fact that o is P*; but also that it might be useful to engage in the pretense that there is such an entity. In order to defend antirealism about facts, two rival realist views are critically examined: the sort of *explanatory realism* that follows from Armstrong's truthmaker argument for states of affairs, and the *simple realism* about facts that follows from the easy approach developed by Amie Thomasson in *Ontology Made Easy* and her previous work on existence questions. After offering a negative assessment of the arguments for realism, we offer some motivations for adopting a fictionalist view of fact talk, and consider how a fictionalist about facts could reply to one of Thomasson's central objections to fictionalism, i.e. how to explain the difference between a (hypothetically successful) assertion that facts exist, and the mere pretense that facts exist.

Keywords Ontology · Fact · Fictionalism · Truthmakers · Easy ontology · Explanatory realism · Simple realism · Application conditions · Amie Thomasson

6.1 Introduction

Some philosophers think that whenever an object *o* instantiates a property *P*, or *n* objects stand in a relation *R*, there exists, as a matter of necessity, a further entity that is not identical to the objects, properties or relations thereby involved, i.e. the

P. Rychter (✉)
Department of Philosophy, Universitat de València, Valencia, Spain
e-mail: pablo.rychter@uv.es

M. Sambrotta (ed.), *Metaphysics Today: In Conversation with Amie Thomasson*, Synthese Library 512, https://doi.org/10.1007/978-3-031-98468-6_6

fact that o is P, or the *fact that those n objects are R-related*. Thus, if Socrates is wise, there exists the *fact* that Socrates is wise—an entity that is not to be identified with either Socrates or the property of being wise. And if you are reading these lines of text, there is the *fact* that you are reading these lines of text, an entity over and above you, the lines of text, and the relation of reading. Such philosophers are *realists* about the existence of facts. Following Thomasson (2015), I distinguish in this paper two sorts of realism about facts: simple realism, and explanatory realism. Simple realism about facts is a position that follows from Thomasson's 'easy approach' to ontology, as applied to the case of facts. According to simple realism, we should believe that facts exist because, as can ascertained by standard conceptual and empirical means, the application conditions of the general term 'fact' are fulfilled. Explanatory realism, on the other hand, is the sort of position that we find, for instance, in Armstrong (2004). According to explanatory realism, we should believe in facts in virtue of an argument to the best explanation: facts are needed to explain different phenomena and, in particular, how truth is grounded in reality. This is Armstrong's *truthmaker argument* for facts, or states of affairs, as he calls them. Against these two sorts of realism about facts, I am interested in defending the plausibility of antirealism. Against explanatory realism, I argue that facts are not really needed in order to explain how truth is grounded in reality: alternative explanations that do not appeal to facts are at least as adequate as standard truthmaker theory. Against simple realism, I argue that the 'easy argument' for the existence of facts can be resisted by the antirealist. In addition to defending the possibility of antirealism about facts, this paper also argues for the plausibility of a *fictionalist* approach to fact talk. Thus even if antirealism is true and so there are no facts, there may be good reasons for preserving philosophical *talk* of facts, albeit understood non-literally.

The structure of the paper is as follows. In Sect. 6.2, I describe the two sorts of realism about facts mentioned above, and some of the arguments that have been presented in their favor. I focus on Armstrong's truthmaker argument for explanatory realism, argue that it is not convincing, and then move on to the 'easy argument' for simple realism about facts. In Sect. 6.3, I take a closer look at this second argument, paying special attention to the issue of how we should state the application conditions of 'fact'. My conclusion there is that the easy argument for simple realism is unsuccessful, because it relies on mistaken assumptions about the application conditions of 'fact'. Having argued for the plausibility of antirealism about facts, I turn, in Sect. 6.4, to defending the plausibility of a fictionalist approach to talk of facts.

6.2 Two Brands of Realism About Facts: Armstrong's 'Explanatory Realism' and the 'Simple Realism' of the Easy Approach

It will useful to start our discussion by considering the following passage from Arianna Betti's recent book *Against Facts*:

> You are reading this book. "A fact!" many would say. A robust piece of reality involving you, reading, and a book. What could be more evident? You most certainly exist, this book most certainly exists, and the relation of reading that relates you and this book most certainly exists. The reading of this book by you: what could be more obvious than that reality in this moment contains such a unified structure? Oh, plenty, I say. **It isn't evident** at all, **nor is it plausible** that there are such things in reality. (Cf. Betti, 2015, Introduction to part I, emphasis added)

In this passage, Betti makes two different claims I agree with. First, she claims that it is not *plausible* that there are facts—things like *you reading this book*. This claim is consistent with both antirealism about facts (the view that there are no facts) and with an agnostic position according to which the arguments for realism are not convincing, and thus not sufficient to move us to accept facts into our ontology. Both antirealism and agnosticism are opposed to realism about facts: the view that reality contains, in addition to you and this book, a *third* thing that is *you reading this book*. The second claim that Betti makes in this paragraph is that the existence of facts is not *evident*: the existence of the fact *you reading of this book* is not a trivial consequence of the true claim that you are reading this book. That this fact exists, and so that facts in general exist, is something that has to be shown by arguments that go beyond simply showing that it is true that you are reading this book. Proving the existence of facts is *harder* than that.

It is in stark opposition to views like this that Thomasson (2007, 2009, 2015) develops her *easy approach*, according to which existence questions, and so, in particular, the question of the existence of facts, can be answered easily by means of conceptual analysis and straightforward empirical observations.

Antirealists and agnostics about facts tend to assume, like Betti, that the ontology of facts is *hard*. Realists, on the other hand, can be sorted into 'explanatory realists', who also think that ontology is hard, and 'simple realists', who assume the easy approach to ontological questions. Realists of both sorts agree that, as a matter of necessity, if you are reading this book, there exists the fact that you are reading this book. But they disagree about what reasons we have for believing that this necessary connection holds: simple realists argue for this belief on the basis of trivial considerations about the meaning of 'fact', whereas explanatory realists rely on indispensability arguments premised on the explanatory work allegedly done by facts. (Note, by the way, that simple and explanatory realists may even agree that the existence of facts is knowable a priori, since the indispensability argument offered by explanatory realists need not rely on empirical considerations.)

In the remainder of this section, I will first consider Armstrong's truthmaker argument for explanatory realism about facts, and then move on to what I take to be the central argument for simple realism. Considering Armstrong's argument will help us to identify a putative theoretical role for facts that will be relevant for our discussion of the fictionalist approach in Sect. 6.4.

According to Armstrong's truthmaker argument (2004: 48), we should accept the existence of facts because this is what best explains how truth is grounded in reality (a phenomenon that requires explanation): truths are true in virtue of having facts as truthmakers. Armstrong's argument is premised on what is known as *truthmaker necessitarianism*—the claim that if T is a truthmaker of *p*, then it is not possible that T exists and *p* is not true. Truthmaker necessitarianism has interesting ontological consequences: Socrates is not a truthmaker of <Socrates is wise>, since it is possible for Socrates to exist while the proposition <Socrates is wise> is false. For similar reasons, Socrates and wisdom taken together do not qualify as a truthmaker of <Socrates is wise>: the mere existence of these two things does not suffice for the truth of the proposition. In contrast, the fact that Socrates is wise, were it to exist, would be a truthmaker for <Socrates is wise>. And since, according to the argument, no other assumption would make for an equally good explanation, we should, in conclusion, accept the existence of facts.

I am not convinced by Armstrong's argument, even though I accept some of its crucial premises. I accept, for instance, Armstrong's assumption that truth is grounded in reality: what is true and what fails to be so depends on how the world is. I also accept the idea that how truth is grounded in reality may require philosophical explanation. That is to say, I accept that truth's being grounded in reality, and how exactly it is grounded, are phenomena to be explained by philosophical theory. However, like many others, I think that there are good explanations of how truth is grounded in reality that do not involve, and hence do not require, the existence of facts (cf. Parsons, 1999, Briggs, 2012, Rychter, 2014). These explanations allow us to reject *truthmaker necessitarianism.* According to the particular view that I favor, that Socrates is wise is true because of a particular feature of reality: that it contains Socrates, a being that actually exemplifies the property of being wise. Of course, Socrates does not necessitate the truth of <Socrates is wise> and so, given the standard definition of 'truthmaker' that involves *truthmaker necessitarianism*, it is not a truthmaker of that truth. Nevertheless, it is part of a perfectly good explanation of why the proposition is true and how its truth is grounded in reality. On this kind of view, *no thing* other than Socrates, and perhaps also the properties and relations that he exemplifies, are necessary for explaining why different propositions about him are true, and, in particular, no *facts* involving Socrates are necessary. All the explanatory work is done by Socrates himself. Furthermore, this explanation is arguably better than the one offered by the realist, since it is more ontologically parsimonious: it does not appeal to any objects (i.e. facts) that we do not have independent reason to believe in.

However, although we need not commit ourselves to the existence of facts in order to explain how truth is grounded in reality, talk of facts may sometimes be useful. When a literally true explanation of the phenomena is too complex, or when we want to compare or discuss the relation between different explanations, it may be useful to pack all the relevant information into fact talk. In such circumstances, we may legitimately pretend that there are facts. Our pretense could follow rules like this: 'Let say that when an object *o* instantiates a property *P*, there exists the *fact that o is P*'. We can furthermore produce putative explanations of truth that appeal to this pretense, by saying, for instance, that the fact that Socrates is wise *makes it true* that Socrates is wise. This is the sort of fictionalist proposal that I will discuss below in Sect. 6.4. According to this proposal, explanatory realists like Armstrong are right that there is an explanatory role to be fulfilled, and that *talk* of facts may be useful for describing the explanatory relations that really hold. The point of disagreement is that, according to the fictionalist, the explanatory role is *not really* fulfilled by facts, but rather by something else.

As mentioned before, explanatory realists about facts, agnostics and antirealists (whether or not they are also fictionalists) typically assume the view that the ontology of facts is hard, in opposition to simple realism and the easy approach to existence questions. On this latter approach, 'existence questions that are fully meaningful and well specified are straightforwardly answerable by making use of our conceptual competence and (often) conducting straightforward empirical enquiries' (Thomasson, 2015: 20). In particular existence questions may often be decided by simply applying the following core rule for the use of 'exists':

E: Ks exist iff the application conditions actually associated with 'K' are fulfilled.

Application conditions are 'basic rules of use that are among those that are meaning-constituting for the term. In the case of sortal terms, these establish certain basic conditions under which the term will succeed or fail in referring' (Thomasson, 2015: 89). Finding out what the application conditions for 'K' are may require some work in conceptual analysis, but once all conceptual issues are settled, there is no further work to do, other than finding out (by standard empirical means) whether the application conditions are satisfied. And if they are, the result is that Ks exist, and so we must be realists about them. This sort of realism, *simple realism*, is not based on considerations about the 'explanatory work' that Ks are supposed to do—only on the idea that the application conditions for 'K' do obtain. In the case of facts, the simple realist says that facts exist because the application conditions of 'fact' are fulfilled. What are these conditions? In order to answer this question, we should not, I think, consider the ordinary meaning and use of the word 'fact', since that is not the meaning that is relevant to our discussion. Rather, we should focus on the special meaning that this word has in the relevant philosophical discourse: we should look at how philosophers introduce and use the word, and how they explicate its meaning. We find a good, interesting example of this in the following passage from Jeffrey King's book on structured propositions:

> Let's call an object possessing a property or n objects standing in an n-place relation, or n properties standing in an n-place relation or etc. a *fact*. We'll call the objects, properties and relations that are parts of a fact its *components*. So the **fact of object o possessing property P** has o and P as components. By definition, all facts obtain. (King, 2007, p. 26, emphasis added for the sake of clarity)

In this passage King is stipulating how he will use the word 'fact', and this stipulation seems to cohere with how the words 'fact' and 'state of affairs' are used by Armstrong and others. We might initially understand this stipulation as implying the following application conditions for 'fact':

(ACF) 'Fact' applies iff some object o or n objects $o_1, o_2, .., o_n$ exemplify some property P or n-ary relation R.

Now, given that these application conditions are clearly fulfilled, the question of whether facts exist has, according to the easy approach, a trivially positive answer. This positive answer, on these particular grounds, is what simple realism about facts amounts to. In other words, simple realism about facts is the position that follows from accepting the following argument:

Easy argument for the existence of facts

P1 E (core rule for 'exists'). Ks exist iff 'K' applies' (i.e. iff the application conditions actually associated with 'K' are fulfilled).

P2 ACF. 'Fact' applies (i.e. the application conditions for 'fact' are fulfilled) iff some object o or n objects $o_1, o_2, .., o_n$ exemplify some property P or n-ary relation R.

P3 Trivial a posteriori claim. Some objects do exemplify properties and relations.

–

C Facts exist

It is crucial to notice here that a supporter of this argument will take ACF, the second premise of the argument, to be a *conceptual truth.* That is to say, she will take it to be a conceptual truth, knowable a priori, that 'fact' applies whenever the right-hand side of ACF is true. It is not built into the argument that ACF is a conceptual truth, but that is how the premise is justified. Actually, that is why the argument deserves to be called an *easy* argument for the existence of facts: all it takes to realize that facts exist is to engage in the sort of *conceptual* analysis that leads to E and ACF, and to accept the trivial a posteriori truth of P3.

In the next section, I will discuss some reasons to doubt the easy argument for the existence of facts and the simple realism that follows from it. Before that, however, it may be interesting to note here how simple realism stands in relation to the dialectics, briefly described above, between explanatory realists and antirealists or agnostics about facts. One reason why antirealists and agnostics think that the explanations of truth offered by realists are not good, or at least not the best ones, is that those explanations are *ontologically unparsimonious*: they incur more ontological commitments than other, similarly good explanations. Now, simple realism by itself does not support the explanatory realist's claim that that facts play a distinctive explanatory role.[1] However, simple realism may be seen here as an ally of explanatory realism, since the claim that facts obviously exist undermines the merits of offering an ontologically parsimonious explanation, as the antirealist claims to do.

6.3 Against Simple Realism about Facts

On my view, the easy argument for the existence of facts presented above can be resisted at its second premise. As I will argue, it is not a *conceptual* truth that 'fact' applies iff some objects instantiate some properties or relations. In other words, it is

[1] Cf. Thomasson (2015: 156): 'Not only does the simple realist not need to appeal to explanatory power or the like to justify her acceptance of the relevant entities, she *cannot* do so. Any attempt to do so would yield only a dormitive virtue explanation.'

conceivable that *o* is *P* and 'fact' fails to apply because there is nothing, other than *o* or *P*, for 'fact' to latch on to. As a result, antirealists and agnostics about facts need not be moved by the easy argument.

We said above that the application conditions of 'fact' might be thought to be captured as follows:

(ACF) 'Fact' applies iff some object *o* or *n* objects o_1, o_2, .., o_n exemplify some property *P* or *n*-ary relation *R*.

And this is certainly what the *easy argument* for the existence of facts assumes. However, it seems to me that King's passage actually presupposes a different condition, something that can be stated as follows:

(ACF*) 'Fact' applies to *x* iff *x* is *some sort of unified structure composed* of an object *o* and a property *P*, or of *n* objects o_1, o_2, .., o_n and a relation *R*, such that *o* exemplifies *P* or o_1, o_2, .., o_n exemplify relation *R*.

Unlike ACF, this second biconditional ACF* points to a relation that must hold between the thing to which 'fact' applies and some other things: objects, properties, and relations. These other things are supposed to *compose* the fact. (The deliberately vague expression 'some sort of unified structure' is meant to imply a particular species of composition, and thus to exclude some composites of *o* and *P*, like their mereological sum, which do not qualify as facts.) This difference between ACF and ACF* has four interrelated implications for our discussion. First, both explanatory realists and antirealists can agree that ACF*, unlike ACF, is the conceptual truth that really captures the meaning of 'fact' in the relevant philosophical discussions. But, second, it is clear that the right-hand side of ACF* is true (and so 'fact' applies) only if there exists an extra thing, *x*, that is not identical to any of its components. And this is precisely what antirealists deny. Thus, as much as an antirealist can accept ACF* as a good elucidation of the philosophical meaning of 'fact', she can also use it in her explanation of why 'fact' has no application. Third, it is clear that, unlike ACF, ACF* is not ontologically ampliative:[2] the left-hand side of ACF* does not introduce any new entity. It rather relabels an entity to which the right-hand side already grants existence. And finally, because of that, ACF* cannot replace ACF as a premise of the easy argument for facts. If we rephrase P2 in terms of ACF*, there is no undisputed claim that could stand in for P3. The claim that '*o* and *P* compose the relevant sort of unified structure', unlike '*o* exemplifies *P*', is clearly not an undisputed claim easily ascertainable by any competent speaker.

I have argued so far that passages like King's, which I take to be representative of the philosophical discussion of facts, seem to suggest application conditions for 'fact' that are more demanding than those presupposed by the easy argument. And

[2] Extending the definition of Chalmers, 2009: 95, let us say that a biconditional is ampliative iff the right-hand side makes an existential claim that is not built into the left-hand side (cf. Thomasson, 2015: 98).

it is these other conditions, captured by ACF*, that amount to a conceptual truth. Now, what if these more demanding conditions were nevertheless satisfied, as would be the case if explanatory realism were true? I'll come back to this issue in a moment. Before that, I would like to broaden the discussion by considering another case to which the previous considerations also seem to apply: corporations. Thomasson (2015, 100) discusses 'corporation', along with other terms for institutional entities like marriages or laws. According to her proposal, the application conditions of 'corporation' could be stated roughly as follows:

(ACC) 'Corporation' applies iff someone filled and signed the relevant papers and paid the relevant fees.

Thus, it seems that we may infer the existence of something new (a corporation) simply from facts about *other* entities: paperwork, fees and signatures. (Let us call these other entities the 'grounding entities'.) However, it seems to me that in order for 'corporation' to apply to something, *that* thing (the corporation) must be appropriately related to the grounding entities, and that this (i.e. the fact that the corporation is related to the grounding entities in a particular way) must be part of the statement of the conditions in which 'corporation' applies. In the case of 'corporation', it seems plausible that part of what speakers know in virtue of being competent with this term is that a corporation is *created* by the signing of the paperwork and the payment of the fees. So it is plausible that the application conditions have something like the following form:

(ACC*) 'Corporation' applies to x iff x is *the causal result* of filling and signing some paperwork, paying the relevant fees, etc.

But, like ACF*, ACC* is not really ontologically ampliative: nothing new is referred to in the left-hand side that is not already present in the right-hand side. And for that reason, ACC* cannot be used in an easy argument for the existence of corporations. Accepting the existence of corporations implies accepting the existence of something *created* by the grounding entities, not only the existence of the grounding entities themselves. Thus, there is space for antirealism about corporations: someone may doubt that there is an x that causally results from filling paperwork and paying fees. Such a position may not be *plausible,* but it cannot be ruled out by simply noting that some paperwork was filled and some fees were paid. In order to rule out antirealism about corporations, we would need the extra claim that there exists something that is the causal result of all of that.

In sum, as I said at the beginning of this section, it is possible to resist the easy argument for the existence of facts by denying that it is a *conceptual* truth that 'fact' applies iff some objects instantiate some properties or relations. And it is equally possible to resist the easy existence of corporations by denying that it is a *conceptual* truth that 'corporation' applies iff someone signed some papers and paid the relevant fees. But consider now the possibility that explanatory realism is true: facts exist and, as a matter of necessity, 'fact' applies iff some objects instantiate some properties or relations. Explanatory realists do not regard this as a mere contingency: they take it to be impossible, albeit conceivable, that there is no fact that *o* is *P* when *o* is *P*. But then, it seems that explanatory realists should take ACF to be true, and the easy argument to be sound. I'd like to make two points in relation to this conclusion. First, even if, on the assumption of explanatory realism, ACF is true, the easy argument no longer deserves its name. As we observed before, the existence of facts is said to be *easy* because it is supposed to be a *conceptual* truth that 'fact' applies iff some objects instantiate some properties or relations. And although the explanatory realist thinks that P2 is a necessary truth, she still rejects the claim that it is a *conceptual* truth. Compare: we may all agree that, as a matter of necessity, 'water' applies iff there is hydrogen and oxygen combined in a certain way. Yet, because this necessary truth is not a conceptual truth, no one will claim that the existence of hydrogen in my glass is easily ascertainable from the realization that there is water in my glass.

The second, related point I would like to make is this: even if ACF and ACC are true on the assumption of explanatory realism, their truth must be explained in more basic terms, and those more basic terms are given by ACF* and ACC* respectively. Thus, the acceptance of ACF and ACC actually relies on a commitment to ACF* and ACC*—and so we may say again that ACF and ACC do not really introduce *new* entities. Let me elaborate on this: implicit in ACF and ACC is the idea that there are *things* to which 'fact' and 'corporation' apply if the conditions in the right-hand side are met. (*Application* is after all a relation between words and things, not just a property of words.) But then we may ask ourselves: what is the relation between the thing to which 'fact' applies and the grounding entities that feature in the right-hand side of ACF? How come that that thing exists (and 'fact' applies to it) whenever the right-hand side of ACF is true? Is it just a coincidence? (Similar questions may of course be asked about the case of corporations.) I think these questions deserve an answer, since refusing to give one would leave us with the sort of mystery that surrounds necessary connections between distinct existences. And I think the natural answers to these questions are provided by ACF*—and by ACC* in the parallel case of corporations. It is not a coincidence that the things to which 'fact' applies exist whenever objects exemplify properties and relations: the former exist because, as ACF* suggests, they are *constituted by* or *made up of* the latter. And it is not a coincidence that a corporation exists whenever someone filled up the relevant papers and paid the fees: for, in doing so, that person *created* (i.e. causally produced) a corporation, as suggested by ACC*. If I am right about this, it seems that ACF and ACC are not self-standing claims: the realist's commitment to them actually relies on the truth of ACF* and ACC*. Thus, those who think that ACC and ACF are true, as explanatory realists do, should not regard them as really ampliative.

Let us sum up the different possible reactions against the easy argument for facts. First, as I hope to have shown, we should in any case reject the *justification* given for the second premise of the argument (ACF): if it is true, it is not a *conceptual* truth. What *is* a conceptual truth is the more demanding biconditional ACF*. Second, antirealists about facts must reject ACF, but they can accept ACF* and deny that the conditions in its right-hand side ever obtain. Third, explanatory realists about facts will accept the truth of ACF and the soundness of the argument. However, because they deny that ACF is a conceptual truth, they will not find the argument at all *easy.* In fact, since ACF is to be explained in terms of ACF*, explanatory realists will find the argument as *hard* as proving the existence of unified structures composed of objects and properties—the sort of thing for which the antirealist demands an argument.

I hope that these conclusions do not rely on the idea (or the rhetoric) that the easy approach to facts performs 'magic', or pulls a rabbit out of hat (cf. Thomasson, 2015: 215). I agree with Thomasson that 'in metaphysics as elsewhere, the only way to pull a rabbit out of a hat is if it's already there'. But, to continue with this picture, my worry in the case of facts is that the rabbit was *not* already there. Or at least that antirealists about facts are entitled to doubt or deny that it is already there, and to demand independent arguments for the claim that it is.

Before closing this section I'd like to say a word about how the words 'thing' and 'object' were used in the arguments above when, for instance, we said that antirealists about facts deny that there exists a *thing* that is composed of objects and properties in the appropriate way, or that hypothetical antirealists about corporations will deny that there is a *thing* that causally results from signing the relevant papers and paying the relevant fees. Thomasson (2007, 2009, 2015) distinguishes three ways in which the words 'object' and 'thing' might be used: (i) as sortal terms 'typically used to track medium-sized, unified, bounded, and independently mobile lumps of stuff' (2015: 196); (ii) in a 'covering sense', according to which they apply whenever some genuine sortal term applies; and finally (iii) in an allegedly 'category-neutral' sense, according to which 'object' and 'thing' could apply independently of whether a specific sortal does. Which of these three uses is involved in claims such as those just quoted? And what consequences follow from that? I think we can safely set aside the first of the three senses distinguished as clearly irrelevant to our discussion. And I suggest that we also leave aside the second option, even though I take it to be an open question whether this use of 'thing' ultimately makes sense or not.[3] That leaves us with the 'covering' sense of 'thing' and 'object'. I see no reason why the antirealist's claims discussed in this section cannot be understood as relying on this sense of the word. The antirealist we have been considering does not reject the existence of facts simply '*on the grounds that* there is no *thing* there' or on the grounds that 'there are not enough objects'.[4] Instead, she rejects the existence of facts on the grounds that no thing satisfies the *real* application conditions of 'fact', which are captured by ACF*, and not by the less demanding ACF, as the easy argument presupposes. Thus, the antirealist that I have pictured in this section is not demanding assurance that a thing exists prior to (or independently of) considering whether the application conditions of 'fact' are satisfied (which would be illegiti-

[3] Thomasson (2007: 6) argues extensively that the allegedly category-neutral use of 'thing' does not make sense. But see Korman (2015: 38) for discussion of those arguments.

[4] Compare: 'on these rules of use for 'thing' [the covering sense], the eliminativist can't say that there is no table in a certain situation *on the grounds that* there is no *thing* there' (Thomasson, 2007: 158). On the worry that 'there are not enough objects', see Thomasson (2015: 218).

mate if 'thing' is used, as we are assuming, in the covering sense). She is rather complaining that the easy ontologist did not succeed in showing that the *real* application conditions of 'fact' are indeed satisfied. In other words: when the antirealist denies that the application conditions of 'fact' are satisfied because no object meets the required specifications, her use of 'object' can be taken as a genuine covering use. Her claim could be taken as short for 'there is no dog, tree, symphony, dragon, witch, fact, etc… that meets the required conditions for being a fact'. (I'll come back to the covering sense of 'object' towards the end of the next section.)

6.4 Fictionalism About Fact Talk

In the previous two sections I defended the plausibility of antirealism about facts. In Sect. 6.3, I argued that *simple realism* about facts (and, more generally, the easy approach to them) relies on a mistaken assumption about which claims about facts are indeed conceptual truths. In Sect. 6.2, I argued that *explanatory realism* about facts is unwarranted: the *truthmaker argument* fails to establish the existence of facts, since the explanatory work that facts are supposed to do can be done equally well by entities of other ontological categories (objects, properties and relations). This makes agnosticism about facts the default position, and leaves room for antirealism. I mentioned in Sect. 6.2, though, that even an antirealist about facts may find it useful to talk *as if* there were facts, in order to describe and explain various phenomena that do not really involve facts. An antirealist can thus be a *fictionalist* about fact talk: she may indulge in such talk, but regard it as part of a useful pretense that conveys information about the phenomena to be explained. (I will call this information the 'real content' of the talk.) Let me illustrate this with a couple of examples.

First, part of the business of orthodox truthmaker theorists consists in determining how exactly truth is grounded in reality. This enquiry takes the form of establishing which generalizations about truthmaking are true. For instance, there is some discussion among truthmaker theorists as to whether the following conjunction thesis (CT) is true or not:

(CT) If e is a truthmaker of $p\&q$, then e is a truthmaker of p and e is a truthmaker of q.[5]

The orthodox truthmaker theorists that I have in mind intend the talk of truthmakers in CT to refer to facts. Thus understood, this principle tells that if a fact makes true a conjunction, it also makes true each of the conjuncts. Determining whether CT and other generalizations like it are true or not is part of what it takes to develop a truthmaker theory, a theory of how truth is grounded in reality. Now the point I want to make is this: even if CT is stated in terms of facts (or, more generally,

[5] Cf. Rodriguez-Pereyra (2006).

in terms of truthmakers), it *need not* be so stated, and the central point it makes (its 'real content') is relevant for anyone who thinks that truth is grounded in reality, whether or not reality contains facts, or any other entity that could be considered a truthmaker. An antirealist about facts, and even an antirealist about truthmakers (someone who holds that nothing fits the job description of truthmakers) may thus learn something from the discussion of whether CT is true or not. She may learn something about how truth is grounded in reality (the real content of CT), even if she cannot take seriously, or literally, the terms in which truthmaker theorists pursue their discussion. Thus, it would be a mistake for her to reject all talk of facts and truthmakers as false and uninformative, just as it would be a mistake for a nominalist to reject all number talk as false and uninformative. It would be better for her to join in with the discussion, making *as if* there are facts (or truthmakers, more generally).[6]

A second example is provided by Jeffrey King's discussion of structured propositions. As we already mentioned, King's theory incurs a commitment to facts, which are his basic building blocks for constructing structured proportions of different sorts. For this reason, an antirealist about facts will have to take King's theory to be literally false. However, it seems to me that such an antirealist would be missing out on an interesting semantic theory because of apparently misplaced ontological worries. It might be argued that when King makes one of the distinctive semantic claims of his theory (like *the proposition expressed by 'John admires Anna' and the proposition expressed by 'Anna is admired by John' are different facts*), his main point (the real content of his claim) is independent of whether there *really* are facts. It certainly seems inappropriate to reject his views about semantics on the grounds that there are no facts. An antirealist or agnostic about facts who wanted to endorse or even assess King's theory, would be better off by considering King's talk of facts to be a *figure of speech*: as talk that we could indulge in without really committing ourselves to the truth of its literal content. In fact, it may even be that King himself uses talk of facts in this way. That would explain his apparent 'carelessness' or lack of concern for the possibility that there are no facts, and thus the possibility that his stipulation about the use of 'fact', discussed above in Sect. 6.3, is unsuccessful. (An alternative explanation of this apparent carelessness would be, of course, that he is assuming something like the easy approach to the existence of facts.)

In short, as these two examples show, an antirealist about facts need not reject all interesting philosophical discussions that are carried out under the presupposition that facts exist. She can join in with those discussions by adopting a fictionalist view of fact talk. That is, she can treat such talk as a useful pretense that conveys important information about certain phenomena (like truthmaking, or compositional semantics) that do not really involve facts. Of course, a fictionalist position like this faces several challenges. Among them, there is the worry expressed by Thomasson (2015: 194), that fictionalism incurs an important debt. I'll consider this argument in what follows, but before that, I'd like to make three general points that I hope will

---

[6] See Rychter (2014: 285) for further discussion.

help us to better understand the sort of fictionalist proposal that I am making, and relate it to the discussion of fictionalism in Chap. 5 of Thomasson (2015).

First, as I understand it, fictionalism is primarily a view about a particular discourse D—not an ontological view about the entities that D is committed to. Thus, it is not by itself an alternative to realism. Take, for instance, fictionalism about mathematical discourse: the view is primarily about the status of that discourse, about how those engaging in it should be understood. As such, the view is not an alternative to platonism, and is in fact compatible with it. The view I have sketched above is primarily about fact talk: about how we should, or could, understand the truthmaker theorist's claims about how facts make propositions true, or King's claims about what sorts of facts structured propositions are. As such, the view is compatible with different stances on the ontological issue of whether facts exist. Relatedly, I think it is also important to properly specify the limits of the target discourse D: what counts as part of it and what does not. Fictionalism about math, for instance, is normally understood as targeting scientific and ordinary talk of numbers, but not philosophical talk. It is thus possible to be a fictionalist about math talk while at the same time holding that platonists and nominalists, arguing in the philosophy room, are to be taken completely literally. In relation to philosophical fact talk, it is possible to distinguish between cases where the existence of facts is at issue (like in the discussion between realists and antirealists about facts) and cases in which facts are taken for granted in order to discuss other issues (how truth is grounded in reality, how sentences manage to express propositions, etc.). The fictionalist proposal that I sketched above is intended to target this second sort of case, and thus it is compatible with thinking that the ontological dispute between the realist and the antirealist about facts should be understood literally.

Second, my proposal is better understood as an instance of *revolutionary* fictionalism, rather than *hermeneutic* fictionalism. According to hermeneutic fictionalism about a discourse D, this discourse is best interpreted as really fictional. According to revolutionary fictionalism about D, we *ought* to use the language of D non-literally, because 'the point of engaging in D would be achieved by pretend-assertions' (Eklund, 2019; see also Kalderon, 2015, 5–7 and Thomasson, 2015: 179). Thus, a hermeneutic fictionalist about mathematical discourse (like Yablo, 2001, 2005) claims that those who ordinarily engage in math talk (including mathematicians) are best understood as not speaking literally. A revolutionary fictionalist about math talk, on the other hand, says that we ought to only use that talk non-literally, since literally understood it is false (because there are no numbers) or unwarranted (since we do not really know yet if there are numbers), and in any case this non-literal use is enough for the purposes that mathematicians and non-philosophers in general pursue with their math talk. With respect to discourse about truthmakers and facts, I prefer not to commit myself to the view that philosophers like Armstrong and other truthmaker theorists are best understood as engaging in fiction when they present their views about truthmaking. All I want to claim for now is that we could understand many of the things they say about facts and truthmakers as non-literal expressions of useful information about how truth is grounded in reality. That is why I am inclined to present my view as a version of *revolutionary* fictionalism.

The third general point I want to make is this: as may already be clear, I do not take my fictionalist proposal about fact talk to be a skeptical view of the ontology of facts. More generally, I do not think of fictionalism, as Thomasson does, as implying a sort of metaontological skepticism, an alternative to the deflationism that follows from her easy approach. According to Thomasson, the fictionalist arrives at the conclusion that traditional ontological debates 'are pointless or misguided, since they are in effect arguing about something that *doesn't matter.* On the fictionalist view, it does not matter whether or not the disputed objects exist, since the discourse does not presuppose this and can get along fine without it'. Thus, for her, 'in many ways the fictionalist approach and the easy approach are similar. Both are equally opposed to both traditional platonism and to traditional nominalism or eliminativism about disputed entities' (2015: 178). However, on my understanding, fictionalism is fully compatible with 'hard ontology',[7] and it actually works as an auxiliary view for the hard ontologist who wants to hold an antirealist or agnostic position about some disputed entities—fictionalism allows her to 'think with the learned and speak with the vulgar'.[8] It allows her to talk as if numbers existed, even if she knows or suspects that there are no numbers. And in the case of facts, it allows her to join in with the discussion that truthmaker theorists pursue about how truth is grounded in reality, without really committing herself to the existence of facts. Notice also that this work of helping the restrained hard ontologist can be done by either hermeneutic or revolutionary fictionalism: if the antirealist is a hermeneutic fictionalist, she can free herself from the charge that her ontological view is incompatible with the target discourse. If she is a revolutionary fictionalist, she can say that the main *point* of that discourse can still be achieved by a non-literal use of it, even if, literally understood, that discourse is incompatible with her ontological view. In any case, we cannot think of this sort of fictionalist as regarding ontological debates as defective, or as 'arguing about something that does not *matter*'. According to this fictionalist, the existence of numbers does not matter to mathematicians (and, in general, to users of mathematical discourse). But it may still matter to ontologists of math, fictionalists included. Likewise, according to this fictionalist, the existence of facts does not matter to philosophers of language like King, or to truthmaker theorists who discuss how truth is grounded in reality (and in general, to users of philosophical fact talk in discussions where the existence of facts is not the issue at stake). But it may still matter to ontologists of facts, fictionalists included. The target discourses 'can get along fine' without the disputed entities, but serious ontology does not.

---

[7] As Eklund (2019) notes in this respect, 'fictionalists are naturally taken to be methodological allies of orthodox Quineans'.

[8] Thus, I do not follow Yablo here, as Thomasson does (2015: 183), in thinking that this sort of motivation for fictionalism (as an aid to antirealism) is 'dead and gone'. I regard Yablo's 'new' motivation (as the best account of the practice) as compatible with, and a necessary complement of the old one.

Having made all these clarifications, let us come back to Thomasson's claim that fictionalists incur a debt. This debt consists in explaining what else must be the case for a claim in D to be literally true, in addition to that claim being appropriate in the relevant game of make-believe (and so for its 'real content' to be true). As Thomasson notes, there is an apparent disanalogy here between the discourses for which fictionalism is advocated (math talk, fact talk, etc.) and overt cases of fictional discourse. Consider, as an example of the latter, Walton's famous case of a group of children pretending that stumps are bears: when one of them says 'there is a bear behind you', it is clear what would have to be the case for the claim to be literally true, in addition to the real content of the claim (*there is a stump behind you*): there must be a bear behind the relevant child. There is a clear difference in this case between the two worldly conditions that correspond, respectively, to the real and literal content of the claim (Thomasson, 2015: 187–188). But this difference is not so clear in the cases that interest us: what else must be the case, in addition to there being more guests than seats, for it to be literally true that 'the number of guests is larger than the number of seats'? And what else is required from the world, over and above Socrates' being wise, for it to be literally true that 'it is a fact that Socrates is wise'? Or, to consider yet another example that Thomasson appeals to in this context, what else must be the case, in addition to the relevant vows and paperwork to have been completed, for it to be literally true that I am married?

On my view, the fictionalist should answer these questions along the following lines: in addition to there being more guests than seats, the literal truth of 'the number of guests is larger than the number of seats' requires the existence of *two objects*, over and above the seats and the guests, such that one of those objects numbers the guests, the other numbers the seats, and the first is larger than the second. The literal truth of 'it is a fact that Socrates is wise' requires the existence of an *object* constituted by Socrates and the property of being wise, and not to be identified with either of those other two things. In these two cases, the fictionalist can point to a difference between the literal content of the claim and what, according to her, is the real content of it in the relevant game of make-believe. Similarly, a hypothetical fictionalist about 'marriage discourse' could say that the literal truth of 'I am married' requires that I *really* have the property of being married, not just that some vows and paperwork have been completed. (However, this last case of fictionalism is on my view implausible for conceptual reasons: it does seem to be a conceptual truth that to be married is just to be a person who completed the relevant vows and paperwork, and so to just pretend that I am married would be a conceptual mistake. In contrast, as argued above in Sect. 6.3, it is not a conceptual truth that a fact exists whenever an object instantiates a property.)

But this is certainly a reply that Thomasson expects:

> There might be some who are tempted to think that more *is* required for there to *really be* numbers, properties, and even marriages and corporations, than can follow via trivial inferences from the truth of the undisputed statement [i.e. the real content of the relevant claims]: namely, that there *really is some (new) object / entity / individual* present. (2015: 195)

And in reply to this, she appeals to the sort of dilemma about the use of 'object' that we already mentioned above at the end of Sect. 6.3. The word has to be being used in either a 'covering sense' or a as sortal term for relatively unified medium-sized lumps of stuff. As we did before in Sect. 6.3, we can set aside this second use of 'object' as irrelevant to our discussion, and focus on the first. (We can also set aside the possibility that the word is being used in an allegedly category-neutral sense.) Can the fictionalist be understood as using the word 'object' in the covering sense when she says that further *objects* are needed for it to be literally true that there exists the fact that Socrates is wise, or for it to be literally true that the number of guests is larger than the number of seats? I think the answer is 'yes', as I will explain in the remainder of this section.

Remember first that, when used in the covering sense, the word 'object' applies if some genuine sortal also does: if 'number', 'fact' or 'corporation' applies to *x*, then 'object' is 'guaranteed to apply' to *x* too. Thus, when 'object' is so understood, we cannot assert or deny that in a given situation there is an object independently of asserting or denying that there is number, fact, corporation, etc.:

> If 'object' is being used in this sense, as a covering term guaranteed to apply if some first-order sortal does, then we cannot try to answer the question of whether there is, for example, a table, a symphony, or a proposition in a given situation by asking whether there is an object. Instead, we can only answer the question 'is there some object?' by way of answering the various sortal-specific existence questions. (2015: 109)

But I think our fictionalist can accept the premise that 'object' applies to *x* only if a specific sortal also does, and also the conclusion that we cannot assert or deny that there is an object without thereby asserting or denying that there is a number, a fact, a corporation, etc. For when our fictionalist says that the literal truth of math talk requires the existence of additional *objects,* the objects she has in mind are indeed *numbers.* So her denial that there are the required objects is not independent of, or somehow prior to, the denial that there are numbers. And in the same way, when a fictionalist about fact talk points to the need for a further object (over and above Socrates and wisdom) to explain what it would take for 'It is a fact that Socrates is wise' to be literally true, she is not using 'object' independently of 'fact'. Her denial that the real content of the claim implies the existence of the required further object *just is* her denial that it implies the existence of a fact.

In sum, if the above reasoning is correct, anyone who adopts the fictionalist proposal that I have outlined in this section can meet the challenge of explaining the difference between a successful pretense that there is a fact and a successful assertion that there is a fact. She can do so by noting that a successful assertion requires the existence of a further *object* that satisfies the conditions for being a fact, since her use of 'object' in this claim can be understood as an instance of the covering use of the word.

6.5 Conclusions

I have discussed different ontological views about the status of facts: explanatory realism, simple realism, agnosticism, and antirealism. My conclusion is that the last two views are plausible, since as I argued in Sect. 6.2, the arguments for explanatory realism are inconclusive and, as I argued in Sect. 6.3, the central argument for simple realism about facts—the 'easy argument'—relies on a mistaken assumption about which claims about facts are indeed conceptual truths. I have also argued, in Sect. 6.4, that agnosticism and antirealism about facts can be fruitfully combined with a kind of fictionalism about fact talk. Fictionalism about fact talk allows the antirealist to join in with philosophical discussions that are carried out in terms of facts, but are actually orthogonal to the existence of facts. My arguments in Sects. 6.3 and 6.4 have pictured antirealists and fictionalists as distinguishing between scenarios in which an object *o* exemplifies a property *P* and no further object exists, and scenarios in which an object *o* exemplifies a property *P* and a further object (a fact) exists. I have argued that, in making these distinctions, antirealists and fictionalists can be understood as using 'object' in its covering sense, and thus need not be understood as intending to use the word in the allegedly category-neutral sense. I'd like to end with a note about the scope of my conclusions: I do not claim that my worries about the easy approach to the existence of facts automatically generalize to other ontological discussions. That is, I have not presented a *general* argument against the easy approach to existence questions, or against simple realism in every other area. In fact, my main worry about the easy argument for facts depends on the details of the meaning of 'fact' in the relevant philosophical discussions. My claim has been that such a meaning is not properly captured by ACF but rather by a more demanding condition, something like ACF*. And although I have extended my argument to the case of corporations, it might be that some sortals have much less demanding application conditions, and that proving the existence of the corresponding entities really is easy. Moreover, and along the same lines, my disagreement with the easy ontologist about facts may end up just being a disagreement about what 'fact' really means, rather than a disagreement about how we should proceed in answering the question of whether facts exist. For all I have said, answering that question may still involve figuring out what the real application conditions of 'fact' are, and then figuring out by standard means whether those conditions are fulfilled or not. It is just that this last step will be *hard* if the application conditions are, as I have argued, more demanding than ACF.[9]

[9] I'd like to thank Amie Thomasson and the participants in the *XXVIII SIUCC* at Granada for very helpful discussion of a previous version of this paper. Financial support leading to this work was provided by the research projects PID2019-106420GA-I00, PID2023-151071NB-I00 and CIPROM/2023/55.

References

Armstrong, D. M. (2004). *Truth and truthmakers*. Cambridge University Press.
Betti, A. (2015). *Against facts*. The MIT Press.
Briggs, R. (2012). Truthmaking without necessitation. *Synthese, 189*(1), 11–28.
Chalmers, D. J. (2009). Ontological anti-realism. In D. J. Chalmers, D. Manley, & R. Wasserman (Eds.), *Metametaphysics: New essays on the foundations of ontology*. Oxford University Press.
Eklund, M. (2019). Fictionalism. In Edward N. Zalta (Ed.), *The Stanford encyclopedia of philosophy* (winter 2019 edition). https://plato.stanford.edu/archives/win2019/entries/fictionalism/
Kalderon, M. E. (2015). *Fictionalism in metaphysics*. Clarendon Press.
King, J. C. (2007). *The nature and structure of content*. Oxford University Press.
Korman, D. Z. (2015). *Objects: Nothing out of the ordinary*. Oxford University Press.
Parsons, J. (1999). There is no 'truthmaker' argument against nominalism. *Australasian Journal of Philosophy, 77*(3), 325–334.
Rodriguez-Pereyra, G. (2006). Truthmaking, entailment, and the conjunction thesis. *Mind, 115*(460), 957–982.
Rychter, P. (2014). Truthmaker theory without truthmakers. *Ratio, 27*(3), 276–290.
Thomasson, A. (2007). *Ordinary objects*. Oxford University Press.
Thomasson, A. (2009). Answerable and unanswerable questions. In D. J. Chalmers, D. Manley, & R. Wasserman (Eds.), *Metametaphysics: New essays on the foundations of ontology*. Oxford University Press.
Thomasson, A. (2015). *Ontology made easy*. Oxford University Press.
Yablo, S. (2001). Go figure: A path through fictionalism. *Midwest Studies in Philosophy, 25*(1), 72–102.
Yablo, S. (2005). The myth of seven. In M. E. Kalderon (Ed.), *Fictionalism in metaphysics* (pp. 88–115). Clarendon Press.

Chapter 7
Ontology Beyond Ramsey's Ladder

Mirco Sambrotta

Abstract In most of her publications, Amie Thomasson develops and defends what she calls a 'easy' approach to ontology, which leads to both a simple ontological realism and a form of metaontological deflationism. In her view, very roughly, the ontological judgment that some first-order claim properly describes worldly facts, or depicts what *really* exists, adds nothing to the first-order claim itself. With the former, one merely reaffirms the latter, without taking on any burden of more substantial explanations. For the latter is not supposed to avert to some deep or substantive criteria that might fail to be met. As a result, ontological questions are easy to answer, and ontological disputes are misguided. The aim of this paper is to offer an account that does justice to ontological discourse. This, though, without bearing the burden of substantial (i.e., represantionalist) standpoints. According to the proposed account, neither questions of that sort are trivial nor are ontological disputes misguided. Instead, ontological judgments serve a significant normative role in shaping and arranging our discursive practice. In particular, they play a distinctive kind of expressive function: that of expressing commitments to the adoption of conceptual structures within which alone descriptions may occur. Ontological discourse thus does not aim to describe, but rather, by showing what should and should not count as reasons or evidence, it articulates the conditions and common ground that make description possible.

7.1 Introduction

Much of language is representational. I consider this a straightforward insight into the relationship between language and the world. Indeed, it seems pretty obvious to consider most of our claims as being in the business of describing how things are,

M. Sambrotta (✉)
Institute of Philosophy, Slovak Academy of Sciences, v.v.i., Bratislava, Slovakia

M. Sambrotta (ed.), *Metaphysics Today: In Conversation with Amie Thomasson*, Synthese Library 512, https://doi.org/10.1007/978-3-031-98468-6_7

representing worldly states of affairs, tracking aspects of the world, stating facts, and so on. It thus seems equally legit to refer to them as the privileged claims that provide the true description of the world; in short, 'factual claims,' 'descriptive statements,' 'declarative sentences,' and so on.

However, this pre-theoretical assumption about the relation of language to the world becomes less trivial when we try to offer a more substantial characterization of the semantic category of fact-stating, which in turn imposes restrictions on the descriptive use of language. For instance, according to naturalism, the only facts are those recognized by natural science.[1] Similarly, physicalism is the view that all facts are ultimately physical facts, and therefore physical discourse is the only autonomous and legitimate factual discourse.

Proverbially, ontology is the study of what there is. If so, it cannot do without the inquiry into what are the facts that comprise the world. Indeed, the intuitive underlying metaontological assumption is that we are ontologically committed to all and only those entities that must exist in order for the descriptive statements we hold to be true to be true. Doing ontology this way, one begins with the factual claims that we take to be true,[2] and then asks what (facts) make them true and hence what (objects, particulars, individuals, etc.) their terms refer to.[3]

For those who operate with 'thick' notions of fact-stating, so that certain linguistic items—sentences, terms—are correctly attributed a semantic substantial property of being factual (e.g., being a statement of natural science), the resulting metaontological conclusion is that there is a general and systematic criterion to answer the questions of ontology (e.g., being endorsed by science).[4] The corresponding ontological programs are then presented as accepting the *real* existence of entities of a certain kind (e.g., members of natural kinds) and, at the same time, as denying the *real* existence of entities of other kinds (e.g., possible worlds, social entities, intentional states, etc.). As a result, since they regard a specific ontological account as more correct than any other, they maintain that the questions posed by ontologists have objective answers. We can call them 'metaontological realists.'[5]

An extreme version of metaontological realism is what Huw Price (2011) calls 'additive monism.' Additive monists hold that all various autonomous discourses

---

[1] It is well-known that for Quine himself, the significant task of the statement-making part of language is that of recording the conclusions of natural science.

[2] Fact-stating utterances are commonly said to be those that are genuinely truth-bearing, or genuinely 'aimed at truth,' or some such.

[3] Following Hale, 'We can argue for the existence of entities of [a disputed] kind by arguing that there are true statements involving expressions of the relevant kind. If, for example, there are true statements incorporating expressions functioning as singular terms, then there are objects of some corresponding kind. If the singular terms are such that, if they have reference at all, they refer to numbers, there are numbers' (2010, 406).

[4] Of course, many different substantive criteria have been proposed and utilized as conditions for what it takes for entities to exist: such as, the so-called 'Eleatic' criterion of having causal power (Armstrong, 1997; Kim, 1992; Merricks, 2001), being mind-independent (Lakoff, 1987; Azzouni, 2004), being observable or trackable (Elder, 1989), etc.

[5] Note that, for instance, ontological anti-realism about numbers that denies that numbers exist can be understood as a form of metaontological realism insofar as it does not deny that the ontological question about the existence of numbers has an objective answer (see Chalmers, 2009, 93).

are factual.[6] Discourses about material, abstract, or mental objects, as well as those about morality, modality, or causation, are all unified by the common concern to depict worldly facts.[7]

Metaontological anti-realists, on the other hand, deny that there is any general and systematic criterion for answering the questions of ontology and, accordingly, deny that there is any objective way to answer such questions.[8] On top of that, they can be viewed as denying that there is a fact-stating role of language of any kind. Or better, metaontological anti-realism can be understood as a form of non-factualism insofar as what it rejects is the assumption that there is a single category of fact-stating discourse or a unique descriptive use of language. Since there is not the property of factuality in any sense, none of our statements can be said to be genuine representational or not. But, if no use of language can be taken as genuinely ontologically committed or not, ontological discourse about what facts (and entities) there really are is simply misguided. Any attempt to raise ontological issues (i.e., are there *really* any such facts? Are there *really* any such entities?) then rests on a category mistake. In other words, once we see that none of our claims is genuinely descriptive, we see that ontological issues rest on a theoretical error about language.

Metaontological anti-realism thus motivates what Huw Price calls 'ontological quietism' (2011, 235–6), which leads us to stop talking about representation altogether and abandon the project of theorizing about word-world relations. It follows that ontological quietism merely 'puts aside' ontology. There are no separate second-order ontological conclusions but simply the business of first-order quantification and ordinary existence assertions: such as, the activities of first-order specialists (e.g., the assertion 'There are about 3500 different species of mosquitoes worldwide' as uttered by a biological scientist or the assertion 'There are four prime numbers less than ten' as uttered by a mathematician), or common-sense existential claims (e.g., 'There is a laptop on my desk').

Consequently, dispensing from word-world representational relations leaves room for a kind of pluralism about the roles of first-order existential quantification. On this view, there is no such thing as a single category of existential discourse, but rather multiple discourses that, despite sharing the same grammatical surface structure, underlie a variety of different functions. Existential talk masks a wide range of

---

[6] For this position, see also Lewis (1983).

[7] Presumably, not the single physical world, for the unifying principle is not that all facts are ultimately physical facts. Instead, they maintain that each autonomous part of language (i.e., framework) stands in representational relation to the relevant part of the metaphysical, rather than physical, world. Accordingly, all the various domains of facts are regarded as sub-domains of a single metaphysical totality. However, it is important to notice that the common concern to portray the facts that comprise the metaphysical totality is understood by additive monists in a substantial way, that is, as a semantic property that underlies the usage of any discourse, *by virtue of which* they turn out to be all descriptive in the same sense.

[8] Again (see footnote 5), noticed that metaontological anti-realism should be distinguished from anti-realism about a certain kind of entities (e.g., anti-realism that denies the existence of numbers): 'somewhat confusingly, this form of anti-realism about numbers is a form of ontological realism about numbers!' (Chalmers, 2009, 93).

different domains (or frameworks, in Carnapian sense), each with its own specific functional role.[9] However, by appealing to the different linguistic roles of the different domains/frameworks, this account does not deny that such utterances are meaningful speech acts but, in any sense, they can be said to be all 'statements of fact' or 'descriptions of how things are.' As a result, it makes no sense to talk about the notion of 'fact' or 'existence' (and, accordingly, 'ontological criterion'); at most, we can talk about different notions of 'fact' and 'existence.'

7.2 Thomasson's Minimalism

Like metaontological anti-realists, Thomasson (2015) denies that the concept of existence is attempting to refer to a general substantive property or relation, the nature of which we can investigate and hope to discover. At the same time, though, she holds that there are objective answers to the questions posed by ontologists. In particular, she adopts a realist view in ontology that accepts the existence of ordinary objects, numbers, properties, propositions, events, and entities of many other sorts. In order to endorse both conclusions without falling into contradiction, the realist view in metaontology she advocates is of a 'minimalist' kind: ontological questions have objective answers, but the answers to those questions are somehow shallow or trivial—in the sense of not requiring substantive philosophical investigation.[10]

Under this minimalist reading, ontological commitment is a more or less trivial consequence of our theoretical commitment, so that ontological practice in a certain sense reflects (or should reflect) our ordinary practice of quantification. First-order existential quantification over entities of a certain kind trivially yields the second-order ontological conclusion—that such entities *really* exist, or that *it's a fact* such entities exist. Unlike the metaontological anti-realist, she thus allows ontological talk, but the latter trivially follows from our existential talk. For this reason, she makes it a trivial matter that ontological claims are truth-apt. Accordingly, she does not need to deny that our existential claims are factual claims. Whereas the metaontological anti-realist requires that our existential claims do not involve ontological commitment (via their existential quantification), for no claim is genuinely descriptive, Thomasson sees all our existential claims as genuinely descriptive. Nevertheless,

[9] As is well known, this view stems from the work of Carnap (1956), who promoted a pluralism of linguistic frameworks, which are distinguished from one another by functional and genealogical factors. This led him to motivate a kind of metaontological anti-realism and ontological quietism. Of course, the idea of 'language games' in the later Wittgenstein (1958) can also be seen in this light.

[10] Thomasson herself (2001) used the expression 'ontological minimalism' to refer to her and Schiffer's (2003) view. In later works, however, she preferred to label her position as 'easy ontology' (2015). A similar approach in philosophy of mathematics is commonly called 'neo-Fregean' (Hale & Wright, 2001). This kind of minimalism also seems compatible with what Campbell (1993) calls 'simple view.'

since the properties of factuality and existence[11] are minimally understood, their notions (as well as that of ontological criterion) are somehow undermined or deflated[12]—and that's why her view in metaontology can also be defined as 'deflationary.'[13]

To sum up, Thomasson concedes that all existential claims are descriptive/factual claims (in this minimal sense). What is withheld is the conclusion that there is a single, substantial category of descriptive discourse (or semantic property of fact-stating), which is supposed to map the bare objective structure of a single independent reality. There is no such function in a substantial sense!

Her minimalist's suggestion must then be that what constitutes the unity of descriptive/fact-stating discourse can only be its quantificational structure. The property of statementhood, the property shared by all factual claim and assertoric utterances, or derived from claimability and assertibility, can only be a grammatical property (a grammatical remark, a grammatical generalisation presented as a substantive proposal). It follows that we should expect no deep (or substantial) conception of factuality and existence. Rather, they turn out to belong to superficial features of language.

Similarly, we should expect no more robust ontological criteria.[14] The only possible metaontological principle, the only possible criterion guiding ontological investigation, can be the formal structure of our existential claims. So to say, ontology and grammar become inseparable. Or better, ontology turns out to be a manifestation of the syntactic skin of all discourses that share the same indicative form (and hence alike in being treated as 'factual'), which though serve a wide variety of different roles. As with pluralists, this leaves room for a multiplicity of (substantial) functions that existential talk can play. At the lower level, there is room for a multiplicity of functions, each associated with different situations that assertions, commitments, and judgments about what exists may play. At the higher order, though, what groups together those very diverse linguistic activities, what they all share, is only their indicative syntax: namely, their quantificational structure.

---

[11] This position somehow reminds me of Frege's well-known view that existence is not a property of individuals, but rather a second-order property: a property (or concept) of concepts.

[12] It must be noticed that the entities we are committed to in this way are not ontologically deflated, or minimal, or have some second-class status, and we should not attribute a difference in ontological standing to them. Rather, Thomasson's view is that the standards for ontological commitment are minimal. As she highlights: «[w] e should not suggest that the entities to which we become committed are in general 'thin and inconsequential', 'ontologically shallow' […] Instead, we should simply say that such entities exist—full stop—and adopt a simple realist view of them» (2015, 146).

[13] Deflationary views in metaontology clearly echo the deflationary theory of truth. For an account of the latter, see mainly Horwich (2010, 299–322). Note, however, that Horwich (1990) uses the term 'minimal truth' more or less as a synonym for the disquotational theory.

[14] Of course, this model still allows for a deflationary conception of the representational relation. The guiding principle is that so long as representationalist intuition trades only in the deflationary notions, minimalists have no reason to reject it.

7.3 Ramsey's Ladder

As mentioned at the beginning of this paper, a common metaontological intuition is that we are ontologically committed to those entities that must exist in order for factual claims we hold to be true to be true. Therefore, one first needs to figure out which claims are genuinely descriptive and only then ask which of them are true and what their terms refer to. As we have seen, there are two ways to take the former inquiry: a 'heavy' way and a 'light' way. On the heavy program, notions such as 'factuality' and 'existence', as well as 'description' and 'representation', are read as standing for substantial properties or relations. For example, according to naturalism, facts are those recognized by natural science. Accordingly, factual claims are scientific claims. We are thereby ontologically committed to the entities to which the names and singular terms occurring in them refer. In the light way, such notions are understood in a deflationary manner. On this deflationary view, we should give up the search for some substantial criterion of factuality or existence telling us what it is for something to be a fact or to exist.

Arguably, the deflationary theory of truth provides a suitable example. According to this theory, truth should not be thought as a special kind of property. Instead, the truth predicate simply serves as a device of generalization (Horwich's, 1990). Demands of a univocal criterion for truth being fulfilled, such as the requirement that truths have truthmakers, are simply rejected (Horwich, 2010, 299–322).[15] Indeed, the idea is that we can move from, say, "Snow is white" to "'Snow is white' is true" without the need for any substantive criteria to be invoked to justify or explain such a move. Following Blackburn (1993), we can call such a maneuver 'moving along Ramsey's ladder': since the latter does not add anything to the former judgment, 'You can as easily make the last judgment as the first — Ramsey's ladder is lying on the ground, horizontal' (78–79). In other words, the move from 'P' to 'It is true that P' does not take us to a new theoretical level.[16]

Although Blackburn's 'quasi realism' mainly concerns moral discourse,[17] Thomasson makes a similar move along Ramey's ladder to justify our right to move

[15] Instead of 'the deflationary theory of truth,' it would be more accurate to speak of 'deflationary theories of truth.' However, all versions agree that 'true' is a redundancy or disquotational notion, the meaning of which is captured in the trivial equivalence schema: "<p > is true iff p" (Horwich, 1998, 103). Consequently, what Bernard Williams called a substantial theory of truth (1973, 202), which hopes to uncover the nature of truth by means of further investigation (such as a correspondence, coherence, verificationist, or other theory of truth,) would be rejected as wrong-headed. So understood, deflationary approaches to truth may stem from the work of Frege, who maintained that: "… the sentence 'I smell the scent of violets' has the same content as the sentence 'it is true that I smell the scent of violets'. So it seems, then, that nothing is added to my thought by my ascribing to it the property of truth" (1918).

[16] Blackburn (1998, 78) also points out that there are philosophies that fail to see that talk of truthmakers, truth-conditions, denotations, and the like adds precisely nothing and 'take advantage of the horizontal nature of Ramsey's ladder to climb it, and then announce a better view from the top.'

[17] It must be noticed that in Blackburn's 'quasi-realism,' moral discourse is not seen as *really* factual; yet, it earns a right and is entitled to the trappings of factuality.

from our ordinary existential claims to ontological conclusions.[18] Construed a la Thomasson, ontological commitment is a more or less trivial consequence of ordinary commitment. As she remarks: "There is a perfectly legitimate deflationary sense of 'fact' in which 'p' is trivially equivalent to 'It is a fact that p'" (2014, 207).[19] As a result, we do not need to meet any deep or substantive criteria to move from 'Xs exist' to 'It is a fact that Xs exist' (or 'Xs *really* exist') and, therefore, conclude that there is a fact about the existence of Xs.[20] So to say, applying Ramsey's ladder to first-order quantification and existential claims (e.g., the activities of first-order specialists, or common-sense existential claims) yields second-order ontological conclusions. But this does not mean we are taken up to a privileged theoretical point. And this is why we can climb this ladder so 'easily.' Or better, 'the reason the climbs are so effortless is that the ladders lead us nowhere' (Price, 2011, 190). Since ontological claims merely reflect the ordinary way of use of our firs-order quantification and existential claims, asserting the former is just another way of asserting the latter.

Ultimately, what Thomasson aims to offer is a kind of 'lightweight' (i.e., 'minimalist' or 'deflationary') analysis of ontological talk, which can explain and make legitimate sense of the realist-sounding discourse without relying on 'heavyweight' ontological program. What she suggests then is that we should simply give up the search for more substantial criteria of factuality (or existence)[21] telling us what it is for something to be a fact or state of affairs (or to exist), much as the deflationists about truth give up the search for the deep nature of truth (2014, 191).[22]

The important point to highlight here is that, according to her, we can accept that the claims concerned are factual and involve ontological commitment without endorsing a more substantial realism that would commit us to use them as explanations for what makes the corresponding first-order quantification and existential claims true.[23] Hence, we can accept the ontological conclusions that there are facts about the existence of the purported entities (e.g., numbers, properties, material objects, possible worlds, intentional states, etc.) or that those entities *really* exist without attempting to understand them as playing any explicative or justificatory role, such as posited as truth-makers for our first-order quantification and existential claims (e.g., mathematical, modal, intentional, or semantic claims). For, again, first-order quantification and existential claims are not made true by second-order ontological claims, but vice versa, the latter are simply derivable by way of trivial

[18] Following Schiffer (2003), she also calls the steps along Ramsey's ladder 'pleonastic transformations' (Thomasson, 2007, 149).

[19] Here Thomasson explicitly quotes Horwich (2006, 45).

[20] Or from 'There are Xs' to 'It is a fact that there are Xs', with the former understood as a first-order existentially quantified claim and the latter as an ontological claim.

[21] Such as having causal powers, being mind-independent, being physical, observable or trackable.

[22] In this respect, it is much like Blackburn's (1993) quasi-realism about the moral and modal, although Thomasson points out that the prefix 'quasi' here can be misleading.

[23] Any more than the claim that the conclusions of natural science correctly represent the world plays no role in explaining or justifying the conclusions of natural science.

inferences from the former without taking on any burden of explanation. Does this favor realism? Yes, but only a realism of a minimal kind.

7.4 Too Easy?

As mentioned above, Thomasson's project has affinities with Blackburn's quasi-realism about moral discourse. According to Blackburn, moral judgments are entitled to the trappings of factuality, although they are not factual in any substantial sense. That is say, moral judgments are not made true by the existence of corresponding moral properties, for the latter plays no explanatory or justificatory role with respect to the former. Thomasson applies a similar strategy to existential discourse. All existential judgments are entitled to the trappings of factuality, leading to the corresponding ontological conclusions, although they are not factual in any substantial sense. That is to say, existential judgments are not made true by the corresponding ontological conclusions, for the latter play no explanatory or justificatory role. The relevant entities are not 'posits' that explain or justify our acceptance of them. In her words: "On the simple realist view, there are the disputed entities all right, but these are not 'posits' that are parts of 'theories', the inclusion of which is justified by their explanatory power" (2015, 157).[24] Therefore, although there is no domain of facts in any substantial sense, and thus no facts in a substantial sense about the existence of any kind of entity, it is quite proper on this view to conclude that there are many—as many as the kinds of entities over which we ordinarily quantify.

But if the usual ontological formulations have no explanatory or justificatory value, what are they for? What function do they serve in the linguistic community? Minimalism deflates the thesis that the various factual bits of language have a common purpose of describing the world, saying how things are, or some such.[25] Accordingly, it deflates the metaontological thesis that all our ordinary existential claims have the common purpose of telling us which entities we have to be ontologically committed to. Instead, the assertive or declarative use of language in general, and the existentially quantified use in particular, underlies a substantial multiplicity of different functions.

In this way, though, it is no longer clear what the different descriptive parts of language are being held to have in common. The minimalist's suggestion seems to be that what constitutes the unity of fact-stating discourse is simply its indicative syntax (e.g., its quantificational structure). As a result, ontological claims turn out to be metalinguistic claims to some extent, serving the function not of describing states of affairs in the world but linguistic habits. This is, of course, a radical

[24] In order to prevent confusion, Thomasson labels her view here and elsewhere (2015, 128–129, 146) as 'simple' realism, as contrasted with 'explanatory' realism.

[25] As mentioned above, the assumption that descriptive discourse is functionally univocal (in a substantial sense) is, on the other hand, endorsed by heavyweight metaontological realists.

conclusion. Wouldn't this view actually be a fundamental challenge to ontology? Wouldn't it make ontology unworthy of being taken seriously?

Since it leaves ontologists uttering the same truths uttered by ordinary people with respect to the existence of numbers, properties, ordinary objects, or whatsoever, it leads giving up much of traditional work on ontology as traditionally understood. There is simply no more room for it! Of course, it is not hard to imagine that some scholars see this more as an advantage than a disadvantage. However, even theorists outside the ontology room might find little satisfaction with a deflationary account of what ontological constructions achieve in language. For the latter, I feel, falls short of doing full justice to our use of ontological vocabulary. A more charitable account would be one that explained the use of this vocabulary in terms of its contribution to our discursive and social practice. One that explained the role the ontological language game plays in communication, in social coordination, and in the linguistic community, in light of the difference in function between first and second-order existence talk. I take for grant that ontological practice is itself a legitimate discursive social practice. Therefore, it would be valuable to provide a perspicuous account of why we do ontology, or rather, why we use ontological vocabulary and constructions.

7.5 Metaontological Expressivism

In order to achieve this goal, the strategy proposed here is to suggest a different understanding of ontological discourse that considers its primary role to be not that of performing a descriptive function of any sort but rather an expressive function. In particular, ontological vocabulary plays a distinctive expressive role with respect to the use of descriptive vocabulary, making explicit features of the discursive framework within which alone description is possible.[26]

With ontological claims, we are not citing facts but determining what should count as a fact. Accordingly, if we take giving reasons (for further claims and other beliefs) as typically a matter of citing facts,[27] with ontological claims, we are not giving reasons but showing what might count as a reason. An ontological judgment then presents a claim as counting as a reason that can occur in justifications or explanations for further claims.

As such, it sets the evidentiary apparatus by which reasons are evaluated. That is, by showing paradigm cases, models of what should and should not count as reasons for claims, ontological judgments in turn determine what should and should not

[26] Of this distinction and relation between expressive role on the one hand and descriptive role on the other, so understood, there is, I suspect, no trace in Thomasson. A plausible and notable precedent might be the discussion of 'hinge propositions' in the latest Wittgenstein (1969).

[27] Following Sellars's inferential functionalism about semantics (1953), for any claim to be contentful it must be able to play a distinctive role in reasoning: it must be able to serve as a reason or evidence for further claims, hence as a premise from which they can be inferred.

count as evidence in support of them. They do not throw arguments or evidence; rather, they help define the very concept of evidence. In a nutshell, ontological judgments present a claim as functioning foundationally, articulating in this way the conditions and common ground necessary for the very activity of giving reasons.

So understood, the central function played by ontological discourse turns out to be that of expressing what we may call 'framework' or 'background' commitments, that is, commitments to the very conceptual structure within which reasons can be given and evaluated, and hence to the very conceptual context in which factual claims can occur.[28]

It follows that an essential element of the expressive power of ontological notation is the capacity to express rulish. Ontological claims are, in the first place, expressions of norms of reasoning: principles that set forth permissions and prohibitions, and in accordance with which to reason. In short, an ontological judgment expresses commitment to a certain way of reasoning and judging. By making it, we are endorsing a certain *epistemic principle* that singles out a certain method of investigation or doxastic practice as the proper and most reliable source of evidence or way of forming beliefs.

However, what guides the activity of reason-giving cannot itself be subject to it. Since ontological claims do not stand in the appropriate relation to reasons, they cannot be in the market of rational considerations. But this is not because the propositions they present cannot be doubted in principle but because they are presented as the bedrock of our very practice of doubt. So to say, they lie at the end of all paths of justification and reason-giving and, following Wittgenstein, 'what lies at the end of all paths of justification and reason-giving are ways of doing—sets of practices learned through training' (1969, § 110; cf. §§ 204, 559; 1958, § 217). What is ungrounded, the 'river bed channeling our stream of thought' (1969, § 167), is not a proposition but a practice.

As a result, since there is no rational basis for their truth, ontological statements cannot be 'true or false' simpliciter. Likewise, since they ultimately articulate what should or should not count as a factual claim, they cannot themselves be factual claims in the first place. Although they look like declarative sentences and assertions claiming truth-evaluable contents (as their surface grammar suggests), they are elucidations of norms governing our discursive practice.[29]

---

[28] 'Hinge commitments', in Wittgenstein's terminology (1969).

[29] It is perhaps worth noting that, in order to avoid the problem of mixed states posed by Geach (1960), a possible solution might be to embrace a hybrid expressivist view, along the lines of contemporary hybrid alternatives to classical expressivism in meta-ethics, which take moral and evaluative sentences to express a mental state that is the resultant of both a cognitive representational state, of the style of beliefs, and a non-cognitive conative state, of the style of desires. Many of such versions of expressivism stem from the work of Allan Gibbard (1986, 1990, 2003), who suggests that normative/evaluative sentences express propositions in an extended sense: not just sets of possible worlds, but sets of norm-world pairs <n,w>. Then, the same normative proposition can be true at the actual world relative to some norms but not others, so that different speakers who agree on all the relevant facts can still evaluate in different ways by employing different norms in making their evaluations. Normative judgments thus express combinations of factual beliefs with systems

So understood, as commands expressed in the indicative mood, they carry normative force. This recognition thus highlights their significant normative role in shaping and arranging rational practices that permeate our lives. It is only within such set of normative practices that argumentations, as well as debates and inquiries, may occur (1969, §§ 162, 167).

But a related (Wittgensteinian) element of this picture is also the idea that only against this background that our argumentative words have their meaning. In this regard, expressing commitment to a certain linguistic framework involves articulating our understanding of first-order judgments and descriptive concepts. Or better, by expressing commitment to a certain linguistic framework, an ontological judgment plays the higher-order function of showing what should count as a correct, meaningful first-order claim. Accordingly, it makes explicit what should count as a genuine singular term (referring to an object in the Aristotelian sense) and what should count as a concept apt to describe the world (i.e., the ones at stake in that claim). An ontological judgment thus introduces a rule into the language such that a particular sentence-type has to be taken as factual. This roughly means that statements involving ontological terms are best understood as having 'the force of prescriptive statements about the use of certain expressions in the object language' (Sellars 1957, §81).[30]

The idea that ontological claims are primarily used to express commitments to the adoption of frameworks by showing how those very linguistic/conceptual schemes are used might suggest that their expressive role is, to some extent, metalinguistic. The contention here, though, is not that by inserting an ontological term, we make it explicit that the utterance is a grammatical remark or a falsifiable hypothesis about language use. Ontological claims are not understood as having the function of tracking or describing linguistic habits. Instead, by claiming them, we are more like endorsing norms. In particular, we are endorsing frameworks within which factual claims can occur, and relative to which they can be rationally evaluated (identifying what is a good reason for what, what is a good reason to doubt, and so on). Neither, however, are those statements to be understood as descriptive reports of commands. They do not report views about how terms and concepts should be used in atomic, first-order predications. Instead, they are expressions of rules. They *show* how terms and concepts are to be used in atomic, first-order predications.[31]

---

of norms (or better, with the acceptance of those systems of norms). Each of these combinations can be captured by a world, representing a way the world might be, together with a second component consisting of a norm, representing a non-cognitive state of norm-acceptance. In this manner, he develops a strategy for combining the relevant non-cognitive attitude (of acceptance towards rules or norms) with belief to generate more complex sentences.

[30] Consider, however, that Sellars is talking here about statements involving modal terms rather than ontological ones.

[31] For an expressivist (non-cognitivist) approach to ontological claims, according to which they express commitments to the adoption of specific discursive frameworks, see also Kraut (2016), which mainly stems from the work of Carnap (1956).

In this picture, which clearly stems from the work of Carnap (1956), ontological disagreements are really inter-framework disagreements, that is, debates about which linguistic frameworks to adopt. In particular, with cross-framework opposition, speakers are expressing opposed states of norm acceptance rather than (or not only) logically contradictory descriptive beliefs. Since they are disagreements about what should count as reason or evidence, they will not be resolved in the way factual disagreements about belief are resolved (i.e., by throwing reasons and evidence). Unlike disagreements within a conceptual framework, they lie beyond the scope of rational resolution.

Interestingly, a disagreement of this sort can be seen as an instance of what is traditionally called 'deep disagreement'. Following Lynch, deep disagreement can be defined as disagreement over basic epistemic principles, where 'a normative principle to the effect that some source or way of forming beliefs has some valuable epistemic status' (2010, 263) is basic when it 'can be justified only by means of an epistemically circular argument' (2010, 265).[32] So understood, deep disagreements turn out to be rationally irresolvable. Since they are clashes of basic reasons for holding a certain principle true (i.e., basic epistemic reasons), there is no further reason to appeal to settle the dispute. It follows that there is no objectively true epistemic principle. Or better, speaking of truth or correctness here does not apply at all.[33]

Nevertheless, an obvious alternative is to appeal to practical reasons: 'What we want is a reason for employing one method over another. That's a practical matter' (Lynch, 2010, 274). Likewise, ontological questions about what really exists (whether material objects, numbers, properties, possible worlds, intentional states, or whatever) ought to be sensibly understood as questions about what we should do: 'Practical questions about whether we should make use of the linguistic forms in question' (Carnap, 1956, 208). Ontological debates then consist in pointing out such comparative advantages and practical considerations about which language as a whole one should choose according to some practical goals.[34]

---

[32] The term was originally introduced by Fogelin, who calls 'deep disagreements' the conflicts between Wittgenstenian hinge propositions or, in Putnam's words, framework propositions (1985, 5–6). For a characterization of deep disagreement as a disagreement that involves hinge propositions, see also Pritchard (2018). Note, however, that this and Lynch's definition can be understood as two sides of the same coin; hence, both equally apply to the characterization of ontological disagreement given here.

[33] Likewise, we do not need to accept, and probably should not accept, that a norm is the best one, that is, the Norm. For a defense of such non-absolutist (and non-factualist) view, see Gibbard (2003) and Field (2009, 2018, 14).

[34] Nevertheless, even though the question of which linguistic frameworks we should adopt is a practical question, it may also be influenced by theoretical considerations (e.g. parsimony, simplicity, fruitfulness, efficiency, etc.). As Carnap (1956, 208) himself acknowledges, in the acceptance of a linguistic framework, '[t]he efficiency, fruitfulness, and simplicity of the use of the thing language may be among the decisive factors. And the questions concerning these qualities are indeed of a theoretical nature'. But which factors are relevant for the decision will ultimately be determined by the practical goals and aims. These are considerations that make it advisable to accept

Somewhat surprisingly, Thomasson seems to endorse a similar view in at least one of her papers (2017). There, she suggests that 'some disputants about existence can, on this model, be plausibly seen as engaged in metalinguistic negotiation not about how a key term should be used, but about whether it should be used at all' (14). Ontological disagreements are thus understood 'not about *how* the term should be used, but rather about *whether* the term should be used at all' (14). Hence, 'they are involved in advocating for different conceptual schemes with different impact on our way of life' (16). If so, though, it is unclear how to reconcile this view with metaontological deflationism (and first-order ontological realism) she advances in most of her works (2007 and 2015, among others). Indeed, if ontological claims trivially follow from true existential claims (and therefore trivially true), they hardly can be understood as expressions of rules carrying normative force. Accordingly, if ontological disputes are easy to settle, they will turn out to be factual debates of some sort rather than metalinguistic negotiations—or non-straightforwardly factual disagreements (Field 2009), as suggested by the normative characterization argued here.

7.6 Concluding Remarks

According to the picture sketched in this paper, in a very Carnapian spirit, ontological discourse aims to go beyond the commitments of local frameworks. Such discourse, though, is best understood as playing an expressive function: that of expressing commitments to the adoption of conceptual structures within which alone description is possible, and hence in which factual claims can occur and be rationally evaluated. In this way, it ultimately turns out to be a useful device that serves to coordinate our socio-linguistic behavior. By understanding ontological claims according to their pragmatic function (i.e., by looking at the practical commitments we undertake in making them as opposed, for example, to their truthmakers), it is possible to make sense of ontological reasoning without having to bear it the burden of representationalism. Indeed, we do not need to bear any substantial characterization of the category or property of existence and factuality. On this view, to say that an ontological account is right because the world is so constituted adds nothing to the claim that a certain basic epistemic principle is right. But to say that a certain basic epistemic principle is right adds nothing to the claim that a certain method of investigation gets the results that it gets. However, this does not mean that the various ontological accounts are all equally right. Or better, to say that the various ontological accounts are all equally right adds nothing to the claim that each is right from its own standpoint, and this could only mean that each account judges as it judges.

---

the relevant tacitly proposed linguistic rules, '[n]ot virtues that give us evidence that the metaphysical view is true' (208).

This view, though, also gains advantages over other non-representationalist approaches, such as minimalism and deflationism. The latter seems to merely disregard and set aside ontological questions as trivial or 'easy' to answer. Thus, they are left in the awkward position of explaining how it is that issues about what *really* exists could be so pervasive and relevant (as they seem to be) in our discursive practice. Unlike such approaches, the present one offers a more conciliatory account that does justice to ontological talk. It is significant to the extent that it is normatively significant. Ontological questions are ultimately questions about how we should arrange our discursive practice. Accordingly, ontological conclusions are ultimately understood as having the significance of rules, hence carrying normative force.

Likewise, this view does not regard disputes about what *really* exists as pointless, shallow, or misguided.[35] Instead, it aims to preserve them by showing why such disagreements are still worth pursuing. They are indeed recognized as serving an important normative role in shaping and arranging the rational practices that permeate our lives.

Does this mean that what exists depends on human agreement on which conceptual framework to use? I would answer this question with Wittgenstein's words: '"So you are saying that human agreement decides what is true and what is false?"— It is what human beings say that is true and false; and they agree in the language they use. This is not agreement in opinions but in form of life' (1958, § 241).

References

Armstrong, D. M. (1997). *A World of States of Affairs*. Cambridge: Cambridge University Press.
Azzouni, J. (2004). *Deflating existential consequence*. Oxford University Press.
Blackburn, S. (1993). *Essays in quasi-realism*. Oxford University Press.
Blackburn, S. (1998). *Ruling passions: A theory of practical reasoning*. Oxford University Press.
Campbell, J. (1993). A simple view of colour. In J. Haldane & C. Wright (Eds.), *Reality, representation and projection*. Oxford University Press.
Carnap, R. (1956). Empiricism, semantics, and ontology. In *Meaning and necessity* (2nd ed., pp. 205–221). University of Chicago Press.
Chalmers, D. (2009). Ontological anti-realism. In D. Chalmers, R. Wasserman, & D. Manley (Eds.), *Metametaphysics: New essays on the foundations of ontology* (pp. 77–129). Oxford University Press.
Elder, C. (1989). Realism, naturalism and culturally generated kinds. *The Philosophical Quarterly, 39*, 425–444.
Field, H. (2009). Epistemology without metaphysics. *Philosophical Studies, 143*(2), 249–290. https://doi.org/10.1007/s11098-009-9338-1
Field, H. (2018). Epistemology from an evaluativist perspective. *Philosophers' Imprint, 18*(12), 1–23.
Fogelin, R. (1985). The logic of deep disagreements. *Informal Logic, 7*(1), 1–8. https://doi.org/10.22329/il.v7i1.2696

---

[35] Neither need we accuse disputants engaged in ontological debates of semantic incompetence—for, following the minimalist approach, if they grasped and used terms properly, they would evidently find the ontological questions *easy* to answer (Thomasson, 2015, 173–174).

Frege, G. (1918). Thoughts. In *Logical investigations* (p. 1977). Blackwell.
Geach, P. T. (1960). Ascriptivism. *Philosophical Review, 69*, 221–225.
Gibbard, A. (1986). An expressivistic theory of normative discourse. *Ethics, 96*(3), 472–485. https://doi.org/10.1086/292770
Gibbard, A. (1990). *Wise choices, apt feelings*. Oxford University Press.
Gibbard, A. (2003). *Thinking how to live*. Harvard University Press.
Hale, B. (2010). The bearable lightness of being. *Axiomathes, 20*, 399–422.
Hale, B., & Wright, C. (2001). *The reason's proper study*. Oxford University Press.
Horwich, P. (1990). *Truth*. Oxford University Press.
Horwich, P. (1998). *Meaning*. Oxford University Press.
Horwich, P. (2006). The nature of meaning. In M. Devitt & R. Hanley (Eds.), *The Blackwell guide to the philosophy of language*. Blackwell.
Horwich, P. (2010). *Truth-meaning-reality*. Clarendon Press.
Kim, J. (1992). Multiple realizability and the metaphysics of reduction. *Philosophy and Phenomenological Research, 52*(1), 1–26. https://doi.org/10.2307/2107741
Kraut, R. (2016). Three carnaps on ontology. In S. Blatti & S. Laponte (Eds.), *Ontology after Carnap* (pp. 31–58). Oxford University Press.
Lakoff, G. (1987). *Women, fire and dangerous things*. University of Chicago Press.
Lewis, D. (1983). General semantics. In *Philosophical Papers* (Vol. I, pp. 198–232). Oxford University Press.
Lynch, M. P. (2010). Epistemic circularity and epistemic incommensurability. In A. Haddock, A. Millar, & D. Pritchard (Eds.), *Social epistemology* (pp. 262–277). Oxford University Press.
Merricks, T. (2001). *Objects and persons*. Oxford University Press.
Price, H. (2011). *Naturalism without mirrors*. Oxford University Press.
Pritchard, D. (2018). Disagreements, of belief and otherwise. In C. R. Johnson (Ed.), *Voicing dissent* (pp. 22–39). Routledge.
Schiffer, S. (2003). *The things we mean*. Oxford University Press.
Sellars, W. (1953). Inference and meaning. *Mind, 62*, 313–338.
Sellars, W. (1957). Counterfactuals, dispositions, and the causal modalities. In H. Feigl, M. Scriven, & G. Maxwell (Eds.), *Minnesota studies in the philosophy of science* (Vol. II). University of Minnesota Press.
Thomasson, A. L. (2001). Ontological minimalism. *American Philosophical Quarterly, 38*(4), 319–331.
Thomasson, A. L. (2007). *Ordinary objects*. Oxford University Press.
Thomasson, A. (2014). Deflationism in semantics and metaphysics. In A. Burgess & B. Sherman (Eds.), *Metasemantics*. Oxford University Press.
Thomasson, A. L. (2015). *Ontology made easy*. Oxford University Press.
Thomasson, A. L. (2017). Metaphysical disputes and metalinguistic negotiation. *Analytic Philosophy, 58*(1), 1–28.
Williams, B. (1973). Consistency and realism. In *Reprinted in problems of the self* (pp. 187–206). Cambridge University Press.
Wittgenstein, L. (1958). *Philosophical investigations*. Blackwell.
Wittgenstein, L. (1969). *On certainty*. Basil Blackwell.

Chapter 8
Modal Metaphysics Made Easy, Not Shallow

Daniel Dohrn

Abstract I critically assess Amy Thomasson's modal normativism. According to modal normativism, modal statements are just object-level statements of semantic rules. My critical assessment has an epistemological and a meta-metaphysical part. As for the epistemological part, Thomasson argues that only normativism can provide a successful modal epistemology. Yet the price is to limit the subject matter of modal claims so as to preclude substantial ones. As for the meta-metaphysical part, the decisive advantage of normativism is that it can account for the incompleteness of the modal profiles of objects like artworks. Yet again the price is that substantial metaphysical inquiry is precluded from the outset. I propose alternatives which preserve the advantage of modal normativism without precluding substantial metaphysical inquiry.

I critically assess Amy Thomasson's modal normativism. While modal normativism has important philosophical advantages, I raise a concern: modal normativism may unduly limit our ambitions of deep metaphysical inquiry. My critical assessment has an epistemological and a meta-metaphysical part.

(1.) I start with summarizing modal normativism: modal statements on metaphysical possibilities and necessities are object-level statements of semantic rules.

(2.) I reconstruct Thomasson's *epistemological* argument for modal normativism: modal normativism is uniquely well-positioned to account for our modal knowledge. I criticize that modal normativism only succeeds in accounting for our modal knowledge by changing the subject matter of such knowledge: instead of concerning substantial metaphysical truths, it boils down to rules of using our language. There may be no substantial modal truths, but this issue should not be

D. Dohrn (✉)
Department of Philosophy, Università degli Studi di Milano, Milan, Italy
e-mail: daniel.dohrn@unimi.it

M. Sambrotta (ed.), *Metaphysics Today: In Conversation with Amie Thomasson*, Synthese Library 512, https://doi.org/10.1007/978-3-031-98468-6_8

decided by epistemology. Substantial modal truths may exist but not be knowable. Yet recent developments in modal knowledge raise prospects that they are knowable.

(3.) In the *meta-metaphysical* part, I (3.1.) elaborate an important advantage of modal normativism. Its therapeutic use can be illustrated by modal truths about artifacts and artworks in particular. There arguably are limits to the metaphysical inquiry into such items. Certain modal questions seem unanswerable. Thomasson makes plausible that we should not bother. Our practice does not only bring items like artifacts and artworks into existence, it also shapes their modal profile. As a consequence, their modal profile may be incomplete as far as our practice leaves it so.

(3.2.) Having pointed out one key advantage of modal normativism, I proceed to the main disadvantage of the doctrine: it precludes deep metaphysical inquiry from the outset, at least as far as modal metaphysics is concerned. While in the case of artifacts and artworks, it seems plausible that our practice determines their modal profile, common sense realism involves the assumption that there are modal profiles of independent things to track. It is not our practice but the independent structure of reality that fixes these modal profiles. I discuss three examples. My first example concerns modal questions of personal identity, which can neither be simply settled by convention nor by empirical inquiry. To be sure, it is to be discussed in how far personal identity may be partially a social construct. My second example is taken from Lewis: could there be a particle that is both positively and negatively charged? It seems that there must be an answer to this question, which we nevertheless cannot settle empirically or by convention. A third example is an argument on the fundamental structure of the world. The examples considered seem to demand deep metaphysical inquiry that goes beyond what is fixed by the conventions that underpin semantic rules.

I use these examples to expose the costs of modal normativism. Normativists must claim that, as far as modal inquiry does not reduce to figuring out empirically informed statements of semantic rules, it is based on a basic misunderstanding of the functioning of modal vocabulary. I argue against such an uncharitable and prohibitive diagnosis.

(3.3.) I close with a reconciliatory proposal. I argue that the main advantages of modal normativism can be preserved without sacrificing substantial metaphysical inquiry. I see two alternatives.

The first alternative is to give up the core claim that modal statements are object-level statements of semantic rules. Instead, there are substantial modal truths about the metaphysical structure of the world. We should still pay due respect to the partially constitutive role of our practice in fixing modal truths. Doing so may lead to leaving the modal profile of things like artifacts and artworks incomplete.

The second alternative is to retain the claim that modal statements are object-level statements of semantic rules, but to add that the latter are not the rules of current language, but the rules that would organize an ideal language. Such a language would best serve to express the ultimate theory of anything emerging at the end of inquiry. This alternative does not save the epistemological argument for normativism, but it would allow us to uphold the idea that ordinary things like artifacts and artworks have incomplete modal profiles.

8.1 Modal Deflationism, Variant Modal Normativism

I shall summarize some key tenets of Thomasson's modal normativism. As I understand it, the view mainly targets metaphysical modality. Modalized claims are used to convey semantic rules for the use of our terms. They are object-level equivalents of explicit metalinguistic rules. For instance, 'bachelors necessarily are male' corresponds to the meta-linguistic rule 'Apply "bachelor" only to males'. An advantage of modal claims compared to explicit meta-linguistic rules is that object-level claims are easier to process. The main advantage compared to non-modalized declaratives is that rules are distinguished from mere fact-stating assertions. A further advantage is that modal claims allow us to convey permissions. One cannot truly say that there is a female US president but only that there could be one. Modal claims are still truth-apt, but they do not need modal facts as truth-makers. They are made true by semantic rules plus empirical facts.

A posteriori claims to necessity like 'water necessarily is H_2O' can be accounted for by factoring them into two parts (Sidelle, 1989; Thomasson, 2021). First, the semantic rules prescribe that if actually *p*, then necessarily so. Second, it has to be empirically settled whether *p*. For instance, the semantic rules for natural kind terms like 'water' determine that the empirically detectable deep structure of the members of the kind they stand for is necessary. If water is H_2O, necessarily so. The template makes the following distinction salient: any transition from a non-modalized statement *p* to a modally qualified statement 'necessarily *p*' or 'possibly *p*' follows entirely from semantic rules. As a consequence, the question: 'assuming that *p*, necessarily so?' is not to be settled by detecting some truth out there but by semantic knowledge. Any justified modal claim can thus be factorized into our knowledge of semantic rules and empirical knowledge about the world.

In the following discussion, I shall avail myself of a distinction which I find difficult to spell out. I shall talk of *substantial* modal truths. I think that this distinction can be formulated *ad hominem* and in this guise should be acceptable to modal normativists like Thomasson. By a substantial modal truth I understand a true modal claim of the form *(metaphysically) possibly, p* or *(metaphysically) necessarily, p*, which is *not* an object-level statement of a semantic rule. Rather it states a metaphysical fact about how things stand entirely independently of our linguistic practices, just as claims like *Socrates is essentially human.* In particular, such a truth cannot be factorized into a non-modal empirical part *p* and a modal part that is entirely due to semantic rules: *assuming that p, necessarily p.*

When I say 'independently of our linguistic practices', I do not want to say that our linguistic practices do not play a role in settling the identity criteria of e.g. Socrates. Yet there is a decisive difference to modal normativism: the latter comes with a hypothesis about the subject matter of modal claims. Such claims do not only involve the application of semantic rules; they state semantic rules.

One potential example of a substantial modal truth to come is

(CHARGE) Necessarily, no particle is both negatively and positively charged.

CHARGE arguably cannot be factorized into the empirical truth *no particle is both negatively and positively charged* and a semantic claim akin to *assuming no particule is both negatively and positively charged, necessarily so* that could be settled on purely semantic grounds. Presumably modal normativists must hold that CHARGE is meaningless.[1] Whereas many metaphysicians assume that there are substantial modal truths, I take the core message of modal normativism to entail that there are none.

8.2 Epistemology

One of the advantages of modal normativism invoked by Thomasson is that it makes modal epistemology easy. As noted by Thomasson, modal epistemology is faced with two challenges:

(i) The Integration Challenge.
(ii) The Reliability Challenge.

(i) The Integration Challenge concerns the integration of metaphysics and epistemology. Any metaphysical theory must at least in principle be compatible with an epistemology telling how we know the theory (Peacocke, 1999, 1). Ditto for *modal* metaphysics and *modal* epistemology.
(ii) The Reliability Challenge is to explain why our capacities of acquiring knowledge in a certain field are reliable. Joshua Schechter divides the challenge into two questions:

> The Operational Question: How does our cognitive mechanism for [modal judgements] work such that it is reliable?
>
> The Etiological Question: How is it that we have a cognitive mechanism for [modal judgements] that is reliable? (Schechter, 2010, 444)

A *naturalistic* version of the Etiological Question for modal epistemology was originally raised by Robert Nozick. Whenever we come up with claims to knowledge of a certain sort, such claims at least in principle have to be compatible with a naturalistic explanation of how we can be reliable in judging these claims, perhaps an evolutionary one. Nozick drew sceptical conclusions for *modal* epistemology from this requirement: '… we do not appear to have such a faculty, and it is implausible that evolutionary processes would instill that within us …' (2001, p. 122).

Thomasson claims that modal normativism fares better than its rivals in meeting the two challenges. Thomasson supports this claim in two ways. Firstly, she argues that other approaches to modal epistemology face huge difficulties or fail outrightly

[1] One may also consider the alternative that CHARGE is false as there is no semantic rule that excludes a particle being positively and negatively charged. Therefore, one might say, there could be such a particle. But I do not see how this alternative fares better than the diagnosis that CHARGE is meaningless with regard to the realist intuition that there is a property out there for our notion of *charge* to track, and that it is not up to us to decide what this property is like.

to meet the challenges. Secondly, she argues that modal normativism does not face these difficulties.

My main concern about this line of argument is closely related to the metaphysical worries I am going to raise, so closely that I am even uncertain whether there is a genuinely epistemological aspect of the problem. Yet since Thomasson herself emphasizes the epistemological advantages of modal normativism, I feel vindicated to address the epistemological aspect separately.

I introduce my concern by what I take to be a widespread picture as to how epistemology proceeds. According to that picture, the metaphysical status of some subject matter is to be settled independently of epistemology, which then inquires into our knowledge of this subject matter. In other words, on the one hand, there is the question of what the subject matter of some claims to knowledge is. On the other hand, there is the question of whether and how we know something about this subject matter. It is one question what the subject matter of modalized claims is, and it is another question whether and how we know them. One may feel that this conflicts with the Integration Challenge. Yet the Integration Challenge is not that one should adapt the subject matter of metaphysics so that one can know it, but that one should not raise claims to knowing metaphysical truths without telling how one can know them.

A weak epistemological argument for modal normativism conforms to the independence requirement just outlined. The argument boils down to supporting the conclusion that an epistemology of modal claims interpreted according to modal normativism easily meets the challenges, and that an epistemology of more substantially interpreted modal claims has difficulties meeting them. If the subject matter of modal claims is as modal normativism has it, we can know them. If it is not, we cannot know them. But that does tell us nothing on what the subject matter of modal claims is. There remains the option that we simply don't know the pertinent modal truths.

Yet there is also a stronger argument: modal normativism is preferable to a more substantial interpretation of modal claims in that it allows us to meet the epistemological challenges. It is this stronger argument I take issue with. The stronger argument violates the intuition that the subject matter of a claim should be determined independently of asking whether and how we know that claim.

One version of the stronger argument is the following:

(i) We have knowledge of metaphysical modalities.
(ii) A necessary condition for knowledge is that one be able to account for it.
(iii) Accounting for knowledge requires to meet the challenges.
(iv) Only modal normativism allows us to meet the challenges.
(v) Hence modal normativism is correct.

Each step of this argument may be resisted.

(i) While Thomasson suggests to reinterpret modal claims such as to make sure that they can be known, there is the relevant alternative to insist that they are substantial, but to avoid a commitment that they are knowable. It may be useful to give an example. Peter van Inwagen is a moderate modal sceptic. He grants that we can know everyday modal claims:

(TABLE) The table over there could have been two feet to the left.

But van Inwagen denies that we are in a position to know far-fetched ones. As examples of far-fetched philosophical claims he presents (van Inwagen, 1998, 67–68):

(PERFECT) It is possible for there to be a perfect being (that is, a being that has all perfections essentially).

(IMMATERIAL) It is possible that I exist and nothing material exists.

(SUFFERING) It is possible that there exist vast amounts of suffering for which there is no explanation.

(COW) There could be a naturally purple cow.

(IRON) There could be transparent iron.

'Possible/Could' here is what van Inwagen calls 'possibility tout court', and others would call metaphysical possibility. Van Inwagen has no doubts that these claims concern substantial metaphysical questions. In contrast, modal normativism commits us to hold that these questions either must be principally decidable by knowing the rules of language plus any relevant empirical knowledge, or that they are meaningless. I shall come back to how plausible this position is in the sections on metaphysics. Here I have used van Inwagen's examples only to illustrate the epistemological alternative of scepticism.

(ii) One may find the requirement to account for our modal knowledge overdemanding. Perhaps it suffices for knowledge to meet a reliability or safety requirement although we cannot account for meeting that requirement.[2]

(iii) One may accept that one has to account for our knowledge but reject the Integration and/or the Reliability challenge as overdemanding or beside the point. Analogously to my concern about premise (ii), there is the significant possibility that one is in a position to acquire substantial metaphysical knowledge without being able to account for that knowledge. One simply lacks access to those features of one's process of acquiring modal knowledge which would allow one to meet the Integration and the Reliability challenge. This does not have to preclude the process from providing substantial modal knowledge. One may acknowledge that the Integration and the Reliability Challenge raise interesting epistemological issues without accepting that our inability to answer them jeopardizes our knowledge claims.

(iv) One may deny that only an epistemology endorsing modal normativism meets the Integration and the Reliability challenge. There is a huge variety of current approaches to modal knowledge of various sorts, which do not subscribe to modal normativism. In particular, a great number of naturalistic or post-rationalist

[2] In defending Williamson's (2007) modal epistemology, Juhani Yli-Vakkuri draws an analogy to sense perception: 'it would be highly implausible to say that you do not know that the patch of grass is green unless you are knowledgeably sensitive to the right wavelengths. Something short of knowledge is enough: e.g., for a reliabilist it suffices that your visual system is reliably sensitive to the appropriate wavelengths.' (Yli-Vakkuri, 2013, 619)

approaches have been forthcoming (Fischer & Leon, 2017). I mention two especially prominent ones. One such approach is Williamson's counterfactual-based anti-exceptionalist epistemology, which has been fiercely criticized (Roca-Royes, 2011; Gregory, 2017) but also defended (e.g. Dohrn, 2020). For instance, building on Williamson, Kroedel gives an interesting evolutionary story how we come to know claims to metaphysical modality by counterfactuals (Kroedel, 2012, 2017). Such an approach seems to perfectly meet the Integration and the Reliability Challenge. A rival approach centres on a naturalistically inspired story how our best science supports empirical knowledge of essences (e.g. Mallozzi, 2021).

Given such approaches to modal epistemology, Thomasson's criticism of selected theories seems somewhat tendentious. She chooses Lewis's (1986) highly contentious modal realism about possible worlds. According to Lewis, possibility claims are made true by worlds. Worlds are causally isolated universes, maximal combinations of spatiotemporally located facts. Any recombination of the actual spatiotemporal mosaic of particular matters of fact is true at some world. Modal realism indeed has sparked epistemological criticism. Here are two representative quotes:

'It seems like black magic that the epistemology of Recombination should just happen to get the metaphysics of possible worlds right, given that there is no connection by which the latter might have informed the former …' (Vetter, 2015, 13).

'Modal realism makes it impossible to see why these principles of recombination, guided by imaginative experiments, should be any good whatsoever at revealing to us what is going on in these independent, spatio-temporally and causally isolated worlds.' (Thomasson, 2021, 2080)

Yet there are also approaches which try to reconcile modal realism with meeting the Integration and the Reliability Challenge (Fischer & Gilbertson, 2019). Thomasson's criticism might be justified if we simply assume a principle of recombination, or if we ground the latter on simply assuming modal realism (as Lewis seems to do). However, things are different if we consider modal realism as an abductive hypothesis how to systematize and explain a pre-existing knowledge of particular modal claims. Such a pre-existing knowledge may be attained for instance by exercises of imaginative recombination. One simply takes some part of the actual mosaic of particular matters of facts and imagines it rearranged. There is plenty of empirical evidence for the usefulness of imagination, which arguably extends to its use in exploring possibilities. That makes an imagination-based approach to substantial modal knowledge a good candidate for meeting the Integration and the Reliability challenge (Dohrn, 2021).

Having discussed the individual steps of the epistemological argument for modal normativism, I shall now come back to my main concern about the overall strategy. The concern is that the epistemological argument for modal normativism tends to trivialize the very point of modal epistemology.[3] That point may be described by strenghening the premise (i) of the strong epistemological argument for normativism

[3] A closely related complaint is made by Fischer: 'Realists about modality offer the most plausible semantics for modal discourse (or so I'll suppose), but they invoke entities with which we can have no causal contact, whether abstracta or causally-isolated concreta. Antirealists can offer straight-

(i) We have knowledge of metaphysical modalities
such as to get
(i*) We have knowledge of *substantial* modal truths.

Epistemology has to account for this knowledge. The epistemology of modal normativism does not account for substantial modal knowledge, but only for our purported knowledge of something else: under the assumption that modal truths are as supposed by modal normativism, the corresponding epistemology accounts for knowledge of these truths. But it does nothing to account for knowledge of substantial modal truths. Thus, it misses the very point of modal epistemology, conceived as a theory of how we know substantial modal truths.

To conclude, it may sound like good news that the epistemology of normativism meets the Integration and the Reliability Challenge, but there is a prize: the modal truths we know are not what we aspired to know. They are not substantial metaphysical truths about the structure of reality. Of course, proponents of modal normativism may say that there is not much lost, or at least that there was nothing substantial to know in the first place. To judge such replies, I shall proceed to the meta-metaphysical part of my critical discussion.

8.3 Meta-Metaphysics

I shall now consider the metaphysics of modal normativism. According to an intuitive robust realism, there are things out there independently of anything our practice contributes to the world. There are also certain modal truths out there. For instance, it is a truth that humans cannot live longer than 130 years or so. Likewise, it is a truth that nothing can travel faster than light speed. Normally, it is assumed that such truths do not concern metaphysical possibility but something more limited, say nomological or some even more restricted possibility. Yet it is also assumed that they are of the same kind as metaphysical possibility, at least as far as they can be characterized by metaphysical possibility plus some limiting condition. For instance, the nomological (im)possibility of p may be characterized by the metaphysical (im) possibility of p & L, where L are the natural laws.

Modal normativism can be introduced by the observation that there are some modal truths which are entirely due to our practices. Assume it is a metaphysical necessity about \$ that 2\$ cannot be valued lower than 1\$ as one can always get 1\$ for 2\$. This necessity is not due to the way the world is, as the impossibility of anything travelling faster than light. It is entirely due to our practice. One may say that it is due to a semantic rule for \$ as a currency.

Other modal truths like the one about traveling at light speed seem of a more mixed character. Modal normativism arguably can also account for these truths. The

forward epistemologies, but they manage this only by analyzing modal claims in ways that undermine much of their interest.' (Fischer, 2017, 5).

semantic rules are built such as to take on board some contribution from the world. For instance, assume it is a natural law that nothing travels faster than light. Then it follows together with the rule for 'natural law' that nothing can (metaphysical 'can') travel faster than light while the natural laws hold.

Now there are certain things for which neither the semantic rules nor empirical facts may settle any modal questions whatsoever that may be asked about them. This brings me to one main advantage and the main disadvantage of metaphysical normativism. I shall start with the former.

8.3.1 Metaphysical Advantages of Modal Normativism

I shall now come to one main advantage of normativism as I see it. Once we understand the role of semantic rules in determining modal truth, we may dissolve spurious problems and define plausible limits for metaphysical disputes:

> Consider questions about, e.g., what conditions precisely it takes for a table to exist (how many particles can be removed while this table still exists?) or what conditions precisely are required for works of music A and B to be identical (exactly how many notes may be changed?). If, as seems plausible, the rules of application and coapplication for our terms simply do not draw any sharp line, then there is no metaphysical fact of the matter to be discovered about when the table would still exist or the works would be identical. Metaphysicians should not then be embarrassed by the failure of their theories to draw any line or by the fact that any attempt to draw one would seem arbitrary. (Thomasson, 2007, 152–153)

Questions like the identity conditions of a table might bother metaphysicians who take modal metaphysics to be in the business of solving substantial riddles about the boundaries of things out there, e.g. figuring out the precise boundaries of a table for different trajectories through spacetime.

Consider modal claims of this sort (let n be some suitably chosen number and T name some particular table:

(TABLE) The maximum number of particles that can be removed while T still exists is n.

I share Thomasson's view that the question whether TABLE might be unanswerable in principle. I feel clueless as to how to answer TABLE in a way that goes beyond the way I feel puzzled by the question, say, how much H_2O atoms exist in the world. In the latter case, I simply shrug shoulders as we cannot empirically settle this question, but so far there is no reason to assume that there is no answer in principle. In contrast, my puzzlement about TABLE runs deeper. There just seems to be no answer out there to whether TABLE.

One explanation is that there simply is no maximum number of particles that can be removed while T still exists. The semantic rules of 'table' plus empirical facts are not sufficiently specific to delimit tables and their properties so precisely. It suffices for them to work sufficiently well in all pragmatic contexts of our dealing with tables. Questions which go beyond these contexts may simply be unanswerable unless we give up our talk about tables and settle for some more specific linguistic practice in the neighbourhood. Modal normativism can perfectly account for this observation.

In contrast, if there were an essence of tables out there independently of our practice, it is tempting to think that such an essence would have to settle whether TABLE. Of course, this temptation may be resisted. Perhaps there are essentially vague objects out there, and the essence of a table is the essence of such a vague object. Alternatively, one may say that there is an answer to TABLE because there is a sharp boundary for the number of particles that can be removed, but we are unable in principle to know it. Still I find the answer given by normativism attractive: our practice together with certain facts determines what a table is by rules for what to call a table, but it simply leaves open how many particles can be removed.

A less pedestrian example that may illustrate the limits of metaphysics are artworks. The identity conditions of artworks have sparked a huge metaphysical debate. Yet arguably there are limits to how far our theory of artworks can detach from our practice of dealing with art:

> So metaphysicians simply can't legitimately purport to discover that, although competent speakers would uniformly apply a painting name such as "the Mona Lisa" to an object in the Louvre that can be hung, restored, or destroyed by fire, really they are wrong to do so, as a painting is really a token performance by an artist in a given place and time… or a type of action … - neither of which can themselves be hung on a wall, restored, or destroyed by fire. (Thomasson, 2007, 153)

I have sympathy for Thomasson's doubts about radically revisionary ontologies of artworks. It seems somewhat plausible that we are not just instrumental in bringing them into the world, but that our practice also has a constitutive role in determining what it is to be an artwork. Part of this constitutive role may be that we cannot go completely wrong in judgements about what kind of thing an artwork is.[4]

I do not feel in a position to rule out a completely opposite view, according to which artworks are part of the independent furniture of the world. For instance, one may share an indicated type theory of musical compositions. According to such a view, the abstract structure of Beethoven's Fifth is out there in Platonic heaven, and all Beethoven did is to point to it by writing down the notes. One may drive such a view further and say that there are substantial truths about what makes an artwork,

---

[4] When I talk of a constitutive role, I am well-aware that this role may be special compared to, say, the role of a constitutional assembly in forming an institution. Eli Hirsch cautions us against saying things like that it depends on our practices what an artwork is, or that if our practices were different, the property of being an artwork would be different:

'…our linguistic conventions determine the meaning of the expression "there exists something"; hence, they determine the meaning of the sentence "there exists something composed of Clinton's nose and the Eiffel Tower". Hence the truth or falsity of this sentence depends in part of our linguistic decisions. It is merely a use-mention confusion to conclude that whether or not there exists something composed of Clinton's nose and the Eiffel tower depends on our linguistic decisions.' (Hirsch, 2011, 70).

I content myself with noting that, when I talk of constitution, I want to cover a role of our practices that does not fall prey to an error of the sort Hirsch has in mind.

which are largely independent of what we call an artwork. Of course, there must be some overlap for our use of 'artwork' to aim at artworks, but we may still go largely wrong in our beliefs about them. Our practice has to track the independent property of being an artwork. Still, I do not find such a view overly plausible to begin with.

If one shares the view that our practices play a role in what it is to be an artwork, one may also be inclined to think that the modal profile of artworks is partly constituted by our practices of dealing with artworks, and even that this profile may be indeterminate with regard to certain questions. For instance, it may be indeterminate how much of the Mona Lisa could be destroyed without the Mona Lisa ceasing to exist.

I have considered two realms of beings for which modal normativism has a certain plausibility: artifacts like tables and artworks like the Mona Lisa. There are modal questions we cannot answer, and it does not seem implausible that they cannot be answered. A promising explanation why they cannot be answered is that the modal profile of the beings at stake is determined by our practice of dealing with these things. As far as our practice leaves open certain modal questions, these questions have no answer. In sum, it is one of the advantages of modal normativism that it can explain why some modal questions simply do not seem to have an answer, even if one were to have any relevant empirical information.

8.3.2 The Problem of Substantial Modal Truths

Having pointed out an advantage of normativism, I shall now come to limitations of the view compared to our metaphysical aspirations. I shall start with the metaphysics of personal identity.

A third example of Thomasson's besides tables and artworks like the Mona Lisa is 'person':

> … a child may learn to apply and refuse the term 'person' by being pointed toward the usual examples and counterexamples, but then be taught in Sunday School that a person is a being with a soul and a spiritual relationship to God. In cases such as these, the metaphysician may legitimately identify a disconnect between the rules and the theory. In such cases the metaphysician may apply a kind of therapy that suggests that the conditions under which the term was actually applied and refused made no appeal to souls; the theological interpretation is a bit of tacked on metaphysics not itself justified by appeal to the rules for using the term. (Thomasson, 2007, 155)

Thomasson uses the example to distinguish bad 'tacked on metaphysics', not backed by the semantic rules for using a term, from the sound metaphysical practice that remains within the boundaries of the semantic rules for terms like 'person'.

Thomasson dismisses Sunday School teachings as a piece of 'tacked on metaphysics', and tacked on metaphysics they may be. I note that, as in the case of artworks, there is also the opposite alternative. Assume that some version of theism is right, and there *is* a soul with a spiritual relationship to God that partly constitutes the human beings we treat as persons. Given these assumptions, our souls are so infinitely important to the success of our lives as described e.g. in Pascal's *Pensées* that our notion of a person may gravitate towards souls even if we are mostly atheists. I shall mention this alternative only to put it aside.

Still there is an intuition that the metaphysics of persons is not simply decided by semantic rules plus empirical results. As contrasted to artifacts and artworks, it is not our practice that forms persons. Persons seem to be independent of our practice in a stronger sense. As an illustration for that intuition, take a *fission* thought experiment (Sattig, 2012). Consider a counterfactual situation in which my brain is split in two and the halves are completed again, two functioning persons resulting. I report three intuitions. First, the situation seems metaphysically possible. Second, there should be a fact of the matter what would happen to me in that situation; whether any of the two persons, and who, would be *me*. Third, I feel at a loss as to what the answer is. It is easy to build a set of possibility claims concerning the case, like the following:

(FISSION) I could undergo fission and be preciscely one of the two resulting persons.

Modal normativism has a ready diagnosis for this case (without being committed to it). There is no answer. The reason is that our rules for 'person' and the first person pronoun plus empirical observation do not provide one.[5] My intuition that there is an answer has to be dismissed as a piece of 'tacked on' metaphysics. Yet this intuition runs deep, or so it seems to me. It is not half-baked Sunday School metaphysics, but it is the kind of intuition that constitutes our very idea of ourselves as part of an independent reality.

I think there is a good response to the intuitions reported that does not jeopardize our notion of an independent reality. Indeed persons, just as artifacts and artworks, may not be as robust furniture of reality as it might seem to common sense realism. Our manifest image of the world with its store of ordinary objects seems to a large extent the result of evolutionary and social processes. When we try to match this image with the teachings of science, ordinary objects seem to rather loosely connect to the beables of science like electrons, fields, and so on, from which they somehow must be built. Given these observations, the claim that the identity conditions of ordinary objects are largely constituted by our conventions does not seem so preposterous. Perhaps persons are among them. Perhaps they more closely resemble artworks and artifacts, and this is why there just is no answer to whether FISSION.

While I admit that this argument has a certain weight in counterbalancing realist intuitions, I doubt that it is of much avail to modal normativism. At some point, we should reach the bedrock of reality. Even if ordinary objects are somewhat loosely

[5] I mention an alternative that one may consider: there is an answer who would be me, but we simply lack some empirical data that are required to give that answer, just as we lack the data for telling the number of H_2O-molecules in the universe. However, I find it very difficult to tell what the relevant data might be. One would expect that our competence with the rules for 'person' and the first person pronoun should at least tell us what the outcomes for different hypothesized data are, just as they tell us what the outcome for 'gold could not have an atomic number n' for different data about the actual atomic number of gold is.

built from the more elementary building blocks of the world, we may go down to these building blocks to formulate our realistic intuitions about a modal profile that goes beyond anything that can be settled by linguistic rules plus empirical data. Here are two examples by David Lewis:

> On still other questions, there seems to be no way at all of fixing our modal opinions, and we just have to confess our irremediable ignorance. I think one question of this kind concerns incompatibility of natural properties. Is it absolutely impossible for one particle to be both positively and negatively charged? Or are the two properties exclusive only under the contingent laws of nature that actually obtain? I do not see how we can make up our minds; or what guarantee we have that there must be some way to settle the question. Certainly we are not entitled just to make the truth be one way or the other by declaration. Whatever the truth may be, it isn't up to us. (Lewis, 1986, 114)

Take

(CHARGE) No particle can be both positively and negatively charged.

Assume that Lewis is right. We cannot answer whether CHARGE. Nor can we simply introduce a semantic rule that settles the answer. That is, no linguistic rule and no empirical data ever tell us the answer on whether CHARGE. Still it seems plausible that there *is* a truth out there.[6]

We may repeat my argument about persons for CHARGE. Perhaps charge (the property of being charged) is like artifacts and artworks in being partly constituted by our practice. 'Charge' is a theoretical term, that may be made up to our convenience. Taking this option seriously, we may insist that we always can introduce a semantic rule. For instance, we may simply use 'charge' so as to exclude that a particle can be both negatively and positively charged. Perhaps we can make do with 'charge', understood in this way. Perhaps there is nothing in what we ever can know about charge that privileges a notion of charge that supports CHARGE over a notion of charge that supports denying CHARGE. Perhaps the consequence is that we may decide the issue either way by convention. But the intuition of realism is that at some point we should definitely reach bedrock: building blocks of the world whose modal profile goes beyond anything that is settled by semantic rules plus empirical data. There is nothing for us to do but track it. Assuming charge is such a building block, we cannot simply decide the issue one way or the other without missing it.

I shall further illustrate the point as follows: 300 or so years ago, people knew nothing of charge. There was no word 'charge' as we use it for an electromagnetic property (I assume that 'charge' is genuinely polysemous). Still there were modal truths about charge. 'Of course', one may say. After all, the modal truth has not been first introduced with the word. It just follows from our way of talking. If we describe

[6] Lewis himself might voice some doubts. When he says 'what guarantee we have that there must be some way to settle the question', he might intimate that we just do not have the epistemic resources of settling the question. Yet he might also be uncertain that there is a modal fact out there. Even if Lewis were to be read in the latter sense, the view expressed in the quote seems incompatible with modal normativism. For modal normativist, there does not seem to be a problem 'to make the truth be one way of the other by declaration', i.e. by introducing a rule for the use of 'charge'.

the situation 300 years ago in our words, we use 'charge' so that there are, and there always have been modal truths about charge, whatever people 300 years ago were able to say or not. Still the rules of a community of speakers 300 years ago did not settle these modal truths. They were as indeterminate as TABLE, we may presume. With regard to charge, our situation may be different now. But it would be highly chauvinistic to think that there are no substantial truths about properties deep down in the fabric of the world for which *we* do not yet have a word. They are 'alien' properties for our language (Lewis, 1994, 475). Are we to say that there are no substantial modal truths about these properties of the sort that CHARGE seems to express? This would fly in the face of the robust commonsense realism I have tried to sketch. Yet whatever the non-modal empirical facts are, the present rules for our words do not seem to settle the modal truths about these properties.

I shall give another example of a deep dispute on the fundamental structure of the world: Jonathan Schaffer (2010, 61–64) famously argued as follows:

Given the world is mereologically structured, whatever complete mereological partition of the whole of the world is fundamental, necessarily so. Since it seems arbitrary that some intermediate partition should be fundamental, either the level of smallest parts or the whole is fundamental. There could be a gunk-world in which there are no smallest parts. Hence necessarily, the whole is fundamental.

This argument strongly depends on modal premises. Nothing guarantees that semantic rules for 'fundamental' and empirical data alone settle the resulting modal issues, given that our everyday use of 'fundamental' does not cover grand metaphysical issues. Independently of whether they do, it seems that any step of Schaffer's argument is either true or false. If not semantic rules plus empirical data, the very structure of reality must make it true or false.

Thomasson discusses a comparable issue, ontological dependence:

> consider John Searle's claim that money depends on collective intentionality, since to be money is to be the sort of thing that is used or regarded as money … This metaphysical claim can be seen as the object-language reflection of a rule of use for the term "money": "Apply 'money' only if 'collective intention' also applies. This rule of use … can be conveyed in the object-language, by using rather than mentioning the terms as: If there is money, there is some collective intention. Or, we may make explicit the status of this latter claim as rule-conveying by adding a modal operator, giving us the claim of metaphysical modality: "Necessarily, if there is money then there is a collective intention" … (Thomasson, 2007, 144)

Thomasson adds:

> This is not to say that the metaphysicians making ontological dependence claims always intend their claims to be understood in these terms—this is not an interpretive claim but rather a proposal for one clear and natural way to understand such claims of metaphysical modality; the onus is on the metaphysician who wants to insist on a different interpretation of the modality in question to say how else the claim should be understood and how she could claim to know it. (Thomasson, 2007, 144)

Thomasson insists that metaphysicians making modal claims about ontological dependence have to specify first in what sense their claims are to be understood, at least if their understanding diverges from the one Thomasson suggests.

I feel that this is too easy. First, 'money' is a tendentious example as it arguably works like 'artwork'. The phenomenon of money depending on our practices, the metaphysics of money may simply fail to go beyond the web of rules, semantic and otherwise, that guide this practice. In contrast, Schaffer's question concerns reality as it is completely independently of any practice.[7]

Second, it seems overly demanding to insist that a metaphysician like Schaffer has to settle meta-philosophical and epistemological questions about the status of modality and its knowability *before* addressing questions of fundamentality. I do not think that the metaphysician is committed to telling how else than according to modal normativism modals should be understood and how one could claim to know them. Just as people may have modal knowledge without being able to answer the Integration and the Reliability challenge, Schaffer may just rely on his competence with modals and on the force of his arguments. This is not to say, of course, that meta-metaphysical and epistemological considerations may not be used to criticize modal debates like the one on fundamentality. In particular, I see a significant possibility that Schaffer's use of 'fundamental' does not have a sufficiently clear meaning for his argument to be truth-evaluable. Yet the problem would not be that there are no semantic rules of the sort: if the empirical facts are thus-and-so, necessarily the whole is fundamental. The problem would already arise for non-modalized claims like 'the universe is the foundation of its parts'. Such claims would lack a clear meaning.

The main concern raised by Thomasson against substantial modal metaphysics is epistemological. There is no naturalistically respectable way of coming to know claims that go beyond those admitted by modal normativism (Thomasson, 2021). The question is in how far this concern supports a theory of modality which diagnoses substantial tenets like Schaffer's with being meaningless or reducible to semantic plus empirical claims. As we have seen, one relevant alternative would be that they are meaningful but that we cannot know them. Compared to scepticism, the diagnosis of a lack of meaning here seems an even more dismal reaction. As far as the epistemological problems are concerned, a more attractive option would be not simply to dismiss a flourishing field of philosophical discourse like fundamentality. Whether such discourses may provide knowledge, and whether we can give a naturally respectable epistemology of them should not be settled from the outset by our theory of modality.

8.3.3 A Reconciliatory Proposal

What are we to do about substantial modal truths which do not follow from our semantic rules and non-modal empirical facts? I see two alternatives of making room for them, which still preserve the normativist insight that some modal profiles might be incomplete.

---

[7]To be sure, Shaffer himself seems to think of metaphysics not in terms of 'writing the book of the world' but as the result of some negotiating process, looking for the best balance of several requirements. Still I guess he would think of his mereological argument as sound.

One alternative would be to give up the radical claim that modal statements are object-level statements of semantic rules. One may still uphold two core claims: many modal truths follow from our semantic rules plus empirical facts, and often the modal profile of objects does not go beyond what follows from our semantic rules plus empirical facts, for instance in the case of artifacts and artworks. Yet one may have to accept that there are substantial modal truths about at least fundamental properties like charge or some even more fundamental properties that our best physics has not yet dreamt of. These truths are not object-level statements of semantic rules, as there are no such rules.

The second alternative would be to adopt a substantially revised version of normativism that detaches semantic rules from the rules *we* are in a position to set up by our linguistic practices and to know. Modal normativism may be formulated for an *ideal* system of semantic rules that emerges at the end of inquiry. The view may be compared to Lewis's best system analysis of laws (Lewis, 1994, 478). Just as in Lewis's analysis the laws are the axioms organizing the ideal physical theory of the world, modal truths à la modal normativism may be object-level statements of the semantic rules that organize the canonical language that best serves to formulate the world view that emerges at the end of inquiry. One may hope these rules together with any empirical facts whatsoever to settle anything about the modal profile of any entity posited, including things like whether CHARGE (assuming charge to be one of the properties countenanced at the end of inquiry). The profiles of high-level ordinary objects may still be incomplete and reduce to what follows from the rules for using ordinary language. Yet of course there may also be more substantial modal truths about them, which follow from the way ordinary objects supervene on deep metaphysical foundations.

To be sure, the epistemological argument for normativism fails for these alternatives. The epistemological challenges re-arise: how can our rules track substantial modal truths? Yet I anyway have voiced my doubts about the epistemological argument. In my view, the more interesting argument concerns incomplete modal profiles of ordinary objects, artifacts, artworks, and so on. Modal normativism well accounts for these incomplete profiles, and so do the two reconciliatory alternatives considered in this section.

Summarizing, I have argued that the purported epistemological advantages of modal normativism are counterbalanced by the burdens of changing the subject matter of modal epistemology. Modal epistemology cannot be made easy without becoming shallow. If we want to preserve knowledge of substantial modal truths, the epistemology of normativism is no option. In a similar vein, modal normativism threatens modal metaphysics with becoming shallow. Its main metaphysical advantage, accounting for the incomplete modal profiles of artifacts, artworks, and ordinary objects more generally, can be realized without sacrificing substantial modal truths.

References

Dohrn, D. (2020). Counterfactuals and non-exceptionalism about modal knowledge. *Erkenntnis, 85*, 1461–1483.
Dohrn, D. (2021). A humean modal epistemology. *Synthese, 199*, 1701–1725
Fischer, B. (2017). *Modal justification via theories*. Springer.
Fischer, B., & Gilbertson, E. (2019). How Lewis can meet the integration challenge. *Journal of Philosophical Research, 44*, 129–144.
Fischer, B., & Leon, F. (Eds.). (2017). *Modal epistemology after rationalism*. Synthese Library.
Gregory, D. (2017). Counterfactual reasoning and knowledge of possibilities. *Philosophical Studies, 174*, 821–835.
Hirsch, E. (2011). *Quantifier variance and realism. Essays in metaontology*. Oxford University Press.
Kroedel, T. (2012). Counterfactuals and the epistemology of modality. *Philosophers' Imprint, 12*, 1–14.
Kroedel, T. (2017). Modal knowledge, evolution, and counterfactuals. In R. W. Fischer & F. Leon (Eds.), *Modal epistemology after rationalism* (pp. 179–196). Synthese Library.
Lewis, D. (1986). *On the plurality of worlds*. Oxford University Press.
Lewis, D. (1994). Humean supervenience debugged. *Mind, 103*, 473–490.
Mallozzi, A. (2021). Putting modal metaphysics first. *Synthese, 198*, 1937–1956.
Nozick, R. (2001). *Invariances: The structure of the objective world*. Harvard University Press.
Peacocke, C. (1999). *Being known*. Oxford University Press.
Roca-Royes, S. (2011). Conceivability and De Re Modal knowledge. *Noûs, 45*, 22–49.
Sattig, T. (2012). The paradox of fission and the ontology of ordinary objects. *Philosophy and Phenomenological Research, 85*, 594–623.
Schaffer, J. (2010). Monism: The priority of the whole. *Philosophical Review, 119*, 31–76.
Schechter, J. (2010). The reliability challenge and the epistemology of logic. *Philosophical Perspectives, 24*, 427–464.
Sidelle, A. (1989). *Necessity, essence, and individuation*. Cornell University Press.
Thomasson, A. (2007). Modal normativism and the method of metaphysics. *Philosophical Topics, 35*, 135–160.
Thomasson, A. (2021). How can we come to know metaphysical modal truths?. *Synthese, 198*, 2077–2106.
Van Inwagen, P. (1998). Modal epistemology. *Philosophical Studies, 92*, 67–84.
Vetter, B. (2015). *Potentiality: From dispositions to modality*. Oxford University Press.
Williamson, T. (2007). *The philosophy of philosophy*. Blackwell.
Yli-Vakkuri, J. (2013). Modal knowledge and counterfactual knowledge. *Philosophical Studies, 162*, 605–623.

References

[illegible]

Chapter 9
Modal Normativism and Modal Representationalism

Theodore D. Locke

Abstract Literal talk of possible worlds and similarity between them is thought to presuppose accepting substantive forms of realism about possible worlds. According to modal normativism, basic modal claims—i.e. claims of metaphysical necessity and possibility—are not descriptive claims reporting on possible worlds, primitive modal facts, or any kind of modal truth maker. Instead, the function of basic modal discourse is normative: basic modal claims express and endorse semantic rules governing the terms involved in those claims. Taking the two idea together, it might be thought that modal normativism should avoid literal talk about possible worlds or should accept a form of fictionalism about possible worlds. In this paper, I will argue that this is false: modal normativism can and ought to keep literal talk of possible worlds and similarity between them. I will show how literal talk of worlds quite naturally falls out the normativist's characterization of the function of modal discourse. Moreover, I will show how literal talk of similarity between worlds naturally falls out of our practice of expressing semantic rules using counterfactual claims on the normativist account.

9.1 Introduction

Literal talk of possible worlds and similarity between them is sometimes thought to presuppose accepting substantive forms of realism about either worlds, abstract entities such as states of affairs, or perhaps some other kind of substantive modal fact.[1] Alternatively, it might be argued that we can talk about possible worlds and

[1] For example see Lewis (1986), Plantinga (1974).

T. D. Locke (✉)
University of Massachusetts, Boston, MA, USA

M. Sambrotta (ed.), *Metaphysics Today: In Conversation with Amie Thomasson*, Synthese Library 512, https://doi.org/10.1007/978-3-031-98468-6_9

similarity between them, but we should do so only as a pretense, i.e. only *as if* such worlds exist.[2] But then we are forced to say that claims about possible worlds and similarity between them are not literally true. According to modal normativism, basic modal claims—i.e. claims of necessity and possibility—are not descriptive claims reporting on possible worlds, primitive modal facts, or any kind of modal truthmaker. Instead, the function of basic modal discourse is normative: basic modal claims express and endorse semantic rules governing the terms involved in those claims but while staying in the object language *using* rather than mentioning the relevant terms.[3] For this reason, it might be thought that modal normativism should avoid literal talk about possible worlds or should accept a form of fictionalism about possible worlds. In this paper, I will argue that this is false: modal normativism can and ought to keep literal talk of possible worlds and similarity between them.[4] An important payoff of this argument is that by characterizing literal talk of possible worlds as primarily a syntactic tool for formalizing basic modal claims, normativism allows us to keep literal talk of possible worlds without the awkward task of fitting substantive worlds or states of affairs into a naturalistic picture of the world.

In Sect. 9.2, I will explain modal representationalism. In order to motivate a normativist account of literal talk of worlds, I will look at a *substantive* form of modal representationalism and raise worries for it. Then I will introduce modal normativism. At first glance, it looks as though modal normativism and modal representationalism are incompatible projects. First, it looks as though modal normativism has no use for literal talk of worlds if basic modal claims serve a normative rather than descriptive function.

Second, it looks as though literal talk of worlds is in tension with the metametaphysical goals of normativism, which is to avoid ontological commitment to entities that are hard to fit into a naturalistic ontology and avoid commitment to claims that do not have a tractable epistemology. In Sect. 9.3, I will argue against this by showing how modal normativism is quite compatible with modal representationalism and showing how literal talk of worlds quite naturally falls out the normativist's characterization of the function of modal discourse. Finally, in Sect. 9.4 I will explain how modal normativism can individuate worlds, and so keep talk of similarity between them, by developing a normativist account of counterfactuals. I will then show how literal talk of similarity between worlds naturally falls out of our practice of expressing semantic rules using counterfactual claims.

Importantly, my main goal in this paper is not to argue that a normativist account of literal talk of worlds is better than other realist or fictionalist accounts, though I will point out important advantages normativism has over these alternatives broadly construed. Rather, my primary goal is to argue that modal normativism and modal

---

[2] Rosen (1990).

[3] Developed in Thomasson (2007a, 2007b, 2009, 2013, 2020).

[4] The arguments I offer here build on those given in Thomasson (2020) and Steinberg (2013) who both argue that worlds can be thought of as pleonastic entities.

representationalism are compatible projects and so normativists should not reject literal talk of worlds but embrace it.

9.2 Modal Normativism and Modal Representationalism

9.2.1 (Substantive) Modal Representationalism

Let *modal representationalism* be a commitment to utilizing literal talk of possible worlds to make claims about what is possible, necessary, and other modal phenomena such as counterfactuals. I will assume that modal representationalism comes with a commitment to the following three equivalences:

R1. *p* is *necessary* if and only if in all possible worlds *p*.
R2. *p* is *possible* if and only if there in some world at which *p*.
R3. *q would* be the case if *p* were the case if and only if there is some *p q* world that is more relevantly similar to the actual world than every *p* non-*q* world.

For a modal represenationalist, all talk of worlds in R1-R3 is literal. By "literal" I mean that the semantic content of an utterance of "*p*" just is [[*p*]]. Compare to a fictionalist utterance of "*p*" where implicit in the semantic content is some kind of operator yielding the content [[F, *p*]] (the content of "according to the fiction F, *p*"). On a literal reading, the semantic content of "there is a world at which *p*" just is [[there is a world at which *p*]], which just is some world at which *p*, i.e. w*p*. On a fictionalist reading, the semantic content of "there is a world at which *p*" is something along the lines of [[Fw, there is a world at which *p*]], which amounts to claiming that according to some world fiction, there is a world at which *p*. Presumably appealing to fictionalism prevents us from being ontologically committed to the existence of worlds by our use of literal world talk. But below I will argue that for modal normativism the practice of making literal claims about worlds and the accompanying ontological commitments is fully explained by the practice of mandating, conveying, and negotiating semantic rules governing non-modal terms.

My explanation contrasts with more substantive explanations about literal talk of worlds. *Substantive modal representationalism* come with two commitments. First, literal talk of worlds tracks the existence of substantive entities. Thus, each of the equivalencies R1-R3 express the *truth conditions* for basic modal claims, i.e. "necessarily, *p*" is true just in case *p* is true in all possible worlds. Even stronger versions of modal representationalism might take worlds to be truthmakers for basic modal claims. However, below I will show that if we take the function of basic modal claims to be non-descriptive then we do not need to posit substantive worlds as entities basic modal claims are attempting to track or as truthmakers for basic modal claims.

The second commitment is a commitment to substantive worlds or some other substantive kind of abstract object such as states of affairs. Let *substantive features of x* be ontologically on par with empirically tractable features of the world discovered by descriptive sciences yet cannot be discovered using only straightforward empirical methods and conceptual analysis. By "ontologically on par" I mean that when we quantify over substantive entities or features we are keeping track of worldly things in the same way we keep track of, say, ordinary material objects when we quantify over them. On a substantive reading of R1-R3, worlds are not just superficially analogous to empirically tractable entities and features but are ontologically on par with those things. However, a key difference is that unlike empirically tractable entities or features such as tables, electrical resistance, cell membranes, or angular momentum, the existence and properties of such worlds cannot be discovered using straightforward empirical methods. This is because such worlds are either actual but non-causal abstract entities or they are some other kind of non-causal entities that are spatiotemporally disjoint from actuality.

Furthermore, since worlds are substantive, our intuitions about what worlds exist or what worlds are possible do not primarily reflect anything about our concepts but instead reflect the existence and nature of substantive worlds or some other abstract object such as states of affairs. Thus, for substantive modal representationalism, there is always a risk that there will be a gap between conceptual facts and facts about worlds. Therefore, for substantive modal representationalism, conceptual analysis can never provide a sufficient explanation of how we know claims about what worlds exist and which worlds are possible.

The first worry for substantive modal representationalism is an explanatory worry. This explanatory worry has an epistemological aspect and a metaphysical aspect. The epistemological worry is that modal representationalism cannot explain the reliability of even simple modal beliefs. The metaphysical worry is that substantive modal representationalism cannot explain why the existence of substantive worlds *make a difference* to even simple modal facts that obtain in the actual world.[5] Assume that substantive possible worlds do in fact exist. According to the substantive realist about worlds, all of the facts about what is or is not possible just are facts about what worlds do and do not exist.[6] Suppose even further that if p is possible then it is necessarily possible and that if p is necessary, then it is necessarily necessary (i.e. assume modal logic S5). Then facts about what is and is not possible are necessary. Then, since we are assuming substantive representationalism, the substantive worlds necessarily exist. But now consider the following counterpossible:

World Annihilation: If all of the substantive possible worlds were to suddenly go out of existence (or they had never existed at all), then we would still know all of the modal facts that we know and, since knowledge is factive, all of the modal facts that we know would still be true.

---

[5] The following argument is a version of a relevance argument found in Salmon (1988) and Blackburn (1993). A similar argument can be found against Platonism about mathematical facts in Liggins (2014).

[6] Most likely in a reductive sense.

This counterpossible seems to be true and indicates that substantive worlds make no difference to certain modal facts and knowledge of those facts. For example, upon the destruction of all the substantive worlds (whether or not we assume actualism), it seems that I would still know that it is metaphysically possible for a red ball to be green rather than red, so it would still be true that it is metaphysically possible for a red ball to be green rather than red.

But if we reject *World Annihilation*, i.e. if we insist that the existence of substantive worlds does make a difference, then I actually wouldn't know that it is metaphysically possible for a red ball to be green rather than red. But this is a highly counterintuitive claim about modal knowledge—surely my knowledge of such a simple claim is reliable. So, even if we grant that substantive worlds exist, their existence seems to make no difference to our actual modal knowledge, and so no difference to the actual modal facts (at least those simple modal facts that reliably fall within the scope of our modal knowledge). Furthermore, even on substantive representationalist views that somehow reduce worlds to some other substantive entities, it is completely mysterious how exactly those entities, say abstract states of affairs, connect with the ordinary obtaining things to metaphysically ensure that that relevant modal facts obtain, e.g., that obtaining samples of water *must* be H_2O or that some actual donkey *might* have written this essay.

The second worry for substantive representationalism is that it fails to provide a reasonable epistemology for our knowledge of what worlds exist and what worlds do not exist. Warren (2016, 2017) provides arguments that raise significant worries for realism about metaphysical structure and non- causal realism more generally. Applying those arguments in this context, we can say that an epistemology about (substantive) possible worlds is reasonable just in case it minimizes errors such that either (a) facts about what (substantive) worlds exist change but our modal beliefs do not or (b) where our modal beliefs reasonably change but facts about what (substantive) worlds exist do not. Above I outlined a worry along the lines of (a), which appealed to a counterpossible. However, we can see the fragility of our modal beliefs on substantive realism by looking at an error along the lines of (b), which doesn't appeal to counterpossibles. Suppose that there are substantive possible worlds where I exist though my biological parents do not. Suppose that without taking any philosophy classes on Kripke I believe that it is possible I could exist though by biological parents do not and so truly believe such worlds exist. But the worry is that I very easily could have taken a class covering Kripke and been reasonably convinced by Kripke's forceful arguments and so no longer believed this is possible. But the facts about what substantive worlds exist would not change. Thus, I would very easily fall into modal error.[7] Further it isn't clear what I could do to recover any kind of stable knowledge about substantive worlds. This means that on the

[7] See Warren (2016) for this argument applied to metaphysical structure and Warren (2017) for the general argument against non-causal realism.

substantive representationalist account, our modal knowledge is not safe since it is largely arbitrary or a matter of luck, i.e. whether our reasonable modal beliefs safely and reliably connect to facts about what substantive worlds exist is a matter of luck.

No doubt there will be rejoinders to these worries for substantive representationalism, but they are still quite forceful. Nonetheless, there is good reason to think that we would *not* do well to eliminate literal talk of worlds from our modal discourse. For example, when I tell someone that it is possible that my biological parents are not who they actually are, my audience might naturally read the claim epistemically, e.g. that for all I know I am adopted. But suppose that is not what I mean, that I am as sure as anyone can be that my biological parents are in fact those people listed on my birth certificate. What I mean to convey isn't something about what I know but about what is *metaphysically* possible. And to get someone to have the intuition I might have had different biological parents, there is a natural reading of, "it is possible that *p*" which means something like *there are ways* the world could have gone such that *p*.[8] That is, there are ways the world could have gone such that my biological parents were not Smith and Jones but Barbara Streisand and Donald Trump, or Queen Elizabeth II and Michael Jordan, or George Eliot and Fyodor Dostoevsky. Furthermore, that is at least three *distinct ways*, or alternatively three *distinct possibilities or possible worlds*, which means that we can count and quantify over ways, possibilities, and worlds.[9]

Finally, when I talk about ways, possibilities, or worlds in which I could have had different biological parents, there doesn't seem to be any implicit pretense in what I am saying, e.g. that I am pretending there are possible worlds. The most natural reading is I mean my claim to be taken literally—there are *literally* at least three different ways I could have had different parents. So, if we want to keep literal talk of worlds, are we forced into accepting substantive modal representationalism and the metaphysical and epistemological worries that come with it? Below I will argue that modal normativism allows us to give a syntactic explanation of modal representationalism and so keep literal talk of worlds without the worries that come with substantive representationalism.[10]

9.2.2 Modal Normativism

According to modal normativism, viewing the function of basic modal claims, e.g. "Necessarily, *p*", as descriptive, quasi-scientific claims attempting to track modal facts or possible worlds "out there" in the same way we use ordinary empirical claims in our attempts to track ordinary entities or properties is misguided. Instead,

[8] See Forbes (1985).

[9] Lewis (1986).

[10] "Syntactic" doesn't mean non-semantic. What I mean is that introducing names for worlds and quantifying over worlds is fully explained by the practice of mandating, conveying, and negotiating semantic rules of use rather than tracking substantive entities.

modal normativists argue the best explanation of the *function* of modal discourse is that modal claims have a primarily *expressive* or *normative* role rather than a descriptive role. Overall, modal normativists argue that:

> "modal language serves the function of mandating, conveying, or renegotiating rules or norms in particularly advantageous ways. For metaphysical discourse…the rules in question are semantic rules (Thomasson, 2020, p. 15)".

Importantly, since basic modal claims are not attempting to describe anything at all, they are *not* attempting to describe semantic rules. Note that in order to describe semantic rules we need to *mention* rather than use the relevant terms. But basic modal claims are generally made in the object language where the terms are *used* and *not* mentioned. So, instead of describing anything, basic modal claims function to mandate, convey, or negotiate the actual semantic rules that govern the use of non-modal terms while using rather than mentioning those terms.[11] Since, on this view, the object language analogues of the semantic rules of use (e.g. "all bachelors are unmarried") are analytic, all modal truths dealing with distinctively metaphysical claims, e.g., claims about identity or persistence, are analytic (e.g. "necessarily all bachelors are unmarried") or derived from analytic truths with an empirical truth (e.g. "necessarily, gold has atomic number 79").[12]

So, according to modal normativism, basic modal claims are object language analogues of rules and permissions that govern the use of non-modal vocabulary:

N1. Statements of "*necessarily*, *p*" express a rule, or consequence of such rule, that governs the use of *p*, e.g. statements of "necessarily, bachelors are unmarried."
N2. Statements of "*possibly*, *p*" express a permission for the use of *p* given the rules that govern the use of *p*, e.g. "possibly, some bachelors are doctors."

Thomasson does not develop an explicit account of metaphysical counterfactuals. But in Sect. 9.4 I will sketch a normativist account of counterfactuals along the following lines:

N3. Statements of "if *p* were the case, then q *would* be the case" express a *semantic requirement* to accept that *q* given *p* as a hypothetical supposition and a range of *relevant auxiliary assumptions* determined by the context, e.g. statements of "if Kim were a bachelor, then Kim would be unmarried."

I will elaborate on N3 below in order to support my thesis that modal normativism is compatible with a syntactic explanation of modal representationalism, i.e. that modal normativism can and ought to keep literal talk of possible worlds and similarity between them.

Before moving on, let me briefly state some advantages to modal normativism, though for a full defense see Thomasson (2020). First, with respect to the

[11] Ibid. See also Thomasson 2007b. Thomasson also notes a number of advantages for expressing the rules in the object language rather than the metalanguage, e.g. the rules might not even be explicitly stateable in a metalanguage to begin with.

[12] See Thomasson (2007a, b: 62–63, 2020).

metaphysics of modality, the advantages are that, since basic modal claims are not descriptive claims, there is no task of trying to find truthmakers for modal claims. Further, there is no trying to explain why, say, the existence of spatiotemporally and causally isolated possible worlds make any difference to what modal claims are true at the actual world. Alternatively, there is no trying to explain how exactly abstract states of affairs or *sui generis* modal properties connect with non-modal properties to metaphysically ensure that, e.g., actual samples of water *must* be H_2O.

Second, since basic modal claims function to mandate, convey, or negotiate semantic rules of use, modal normativism provides a clear and plausible story for how we come to know basic modal claims made in metaphysics: we come to know basic modal truths only through conceptual analysis and by making straightforward empirical discoveries. For example, coming to know that *gold necessarily has atomic number 79* does not require that we have some insight into substantive modal properties of gold. Suppose there is a rule for 'gold', stated in a metalanguage, along the lines of.

> Apply 'gold' to some sample of a yellow, malleable precious metal iff that sample has atomic composition *C* (where *C* is determined by the atomic composition of whatever sample was used in the actual baptism of the term 'gold').

According to modal normativism, this rule, along with the straightforward empirical discovery that the atomic composition of the relevant samples in our environment is having 79 protons, is expressed with the *de re* modal claim "gold necessarily has atomic number 79."

However, nothing in my discussion of modal normativism so far suggests anything about what normativists should say about literal talk of worlds.[13] So far I've only noted that basic modal claims do not track or describe worlds and do not need as worlds as truthmakers. At this point, one option is to either eliminate talk of possible (or impossible) worlds altogether or to keep talk of possible (or impossible) worlds but interpret it along fictionalist lines rather than take that talk literally. I will now argue that either option is unnecessary by showing how a syntactic explanation of modal representationalism naturally falls out of modal normativism.

9.3 Worlds for Modal Normativists

9.3.1 The Insights of Normativism and Representationalism

Normativism and representationalism provide two important insights for thinking about modality. The first insight comes from normativism. Because the function of modal language is normative, we do not need to treat modal statements as *descriptive* statements about entities "out there", and so we do not need to posit the existence of substantive modal properties or substantive worlds to make modal claims

[13] Though Thomasson (2020: 132–137) begins to develop the idea.

true. This in turn allows us to use basic modal language without taking on substantive metaphysical commitments, which in turn gives us a clear and plausible picture about what we do with basic modal claims and how we come to know basic modal claims—through straightforward empirical methods and conceptual analysis.

The second insight comes from representationalism. As suggested above, literal talk of *worlds* or *ways* things could have been nicely captures a distinction between metaphysical and epistemic modalities. Further, literal talk of worlds or ways things could have been allows us to make more fine grained modal claims than we can with basic modal claims, e.g. not only is it possible I could have had different biological parents, there are many different ways I could have had different biological parents. Finally, by talking about worlds and similarity, we have another way of expressing counterfactual relations. For example, suppose I am a social scientist and claim "if the United States were to adopt a universal basic income, then racism and sexism would still be a problem." I might clarify and restate this claim by asking someone to consider a possible world where, counter to fact, the United States adopts a universal basic income but where all of the facts about race and gender relations are held fixed. Then, after a proper elaboration of those facts, someone might agree that in such a possible world, racism and sexism is still a problem in the United States.

The *prima facie* problem is that it looks difficult to retain the insights of both normativism and representationalism. On the one hand, without talk of worlds, as well as talk about similarity between them, we lose a powerful framework for making even more fine-grained modal claims than basic modal vocabulary will allow.[14] On the other hand, literal talk of worlds and similarity seems to raise the ontological and epistemological worries discussed above, and it would be best to avoid these worries if we can. Fortunately, it is easy to avoid these worries by giving a normativist explanation of talk of worlds and metaphysical similarity.

9.3.2 The General Argument

The general argument is actually quite simple.[15] First, assume the normativist interpretation of basic modal claims outlined in Sect. 9.2.2. Now grant the extensional relationship between basic modal claims and talk about worlds given by R1-R3. According to modal normativism, basic modal claims mandate or convey semantic rules but while using rather than mentioning the relevant terms. Given the extensional relationship between basic modal claims and talk about worlds, this means that talk about worlds can also be used to mandate or convey semantic rules but while using rather than mentioning the relevant terms. Assuming normativism, by N1, statements of "necessarily, *p*" express a rule, or consequence of such rule, that

[14] See Lewis (1973, 1986).

[15] I offer this argument in Locke (2019) and similar arguments are offered in Thomasson (2020: 133). In Locke (2023) I look more closely at the explanatory connections between modal normativism and possible worlds semantics.

governs the use of *p*. For example, statements of "necessarily, if a clay statue is flattened, then the statue is destroyed" conveys a rule governing proper use of the sortal "statue". By the equivalence given by R1, this means that statements of "in all possible worlds *p*" can also be used to express a rule, or consequence of a rule, that governs the use of *p*. Of course, in ordinary language, we often speak in terms of "ways" rather than "worlds", e.g. "for every *way* that you might flatten a clay statue, the statue will be destroyed but not the clay." Thus, we are able to keep the metaphysical and epistemological insights of modal normativism but still keep talk of worlds to make more fine-grained modal claims.

Nonetheless, more needs to be said about how this general argument works. Why should the normativist grant the extensional relationship between basic modal claims and talk about worlds given by R1-R3? One reason is that, on the normativist account of modality, there is no metaphysical cost that comes with literal talk of worlds because worlds are not doing any serious explanatory work in our account of modality (or in our accounts of content). We neither need worlds to serve as truth-makers for basic modal claims nor to secure proper semantic values for basic modal claims, e.g., serve as semantic magnets that explain the meaning of basic modal claims. This is because the function of basic modal claims is to mandate and convey norms and not track descriptive features. This leaves open the following explanation: literal talk of worlds merely provides a different syntactic framework for offering alternative illustrations of semantic rules, or consequences of those rules, that govern the use of non- modal vocabulary.

Basically, on the normativist account of modality, R1-R3 are conceptual truths that draw a connection between seemingly controversial entities, i.e. worlds, and less controversial claims involving basic modal operators.[16] But since R1-R3 are conceptual truths, there is nothing controversial about talk of worlds, ways, or possibilities. Steinberg (2013) and Thomasson (2020) also suggest that when read from left to right R1-R2 express conceptual truths. But I am arguing that R1-R2 (and R3) are also conceptual truths when read from right to left. Further, while Steinberg suggests that conceptual truths are those knowable "simply on the basis of mastery of the relevant concepts (770)", not much additional explanation is given for why we should take R1-R2 (or R3) to be conceptual truths. I will now attempt to flesh out this explanation by drawing on Thomasson's recent work on linguistic function and the work of functional linguists.[17]

In short, the reason that R1-R3 are conceptual truths is that talk of worlds, ways, and possibilities are mere syntactic variations of basic modal claims. Since (i) the meaning of basic modal claims is fully explained by their function of conveying, mandating, and negotiating semantic rules, and (ii) talk of worlds is merely a syntactic variation of basic modal claims, then (3) the meaning of claims involving

[16] See also Hale and Wright (2001), Thomasson (2007a, b, 2014, 2020), Steinberg (2013), Locke (2018, 2019, 2020 2023).

[17] Thomasson (2024), Halliday (2009).

possible worlds is also fully explained by their function of conveying, mandating, and negotiating semantic rules. There are two clarifications that need to be made.

First, mere talk of "syntactic variation" isn't sufficient since the claim is not that we can replace "it is possible that…" with "there is a way that…" in the same way we can replace the former with "es posible que…" It is better to say that talk of worlds, ways, and possibilities are *syntactic metaphors* of basic modal claims. However, a syntactic *metaphor* is not a metaphor in the normal sense of the term where a metaphor is a figure of speech that often relies on comparisons across rhetorical categories, e.g. comparing a mercurial temperament to stormy weather. Rather, a *syntactic* metaphor involves a shift across syntactic categories. It is more like a derivational morpheme where a new expression is formed from an existing lexical expression such that the syntactic category changes (but not necessarily the meaning). For example, rather than speaking about *bathing*, which involves the verb "bathe," we can speak about *taking a bath*, which involves the noun "bath." Or rather than using the verb phrase "washing the car" we might instead use the noun phrase "a car wash."[18] Semantically, syntactic metaphors more or less amount to the same thing as the original expression. But syntactic metaphors often give us different forms of expressive power. For example, rather than saying "Smith commented on my talk" I can say "Smith gave me many helpful comments", which can be followed up with quantificational questions such as "How many comments did Smith make?" I am arguing that, for the modal normativist, talk of worlds, ways, and possibilities is just a syntactic metaphor for talk about what is possible and necessary. Below, I will show how this in turn gives us different ways of conveying and mandating semantic rules and consequences of those rules.

The second clarification is that another reason that R1-R3 are conceptual truths is that the meaning of each component has the exact same metasemantic explanation—the introduction and elimination of these claims is fully explained by their non-descriptive function. Consider that, according to normativism, the meaning of basic modal operators is fully explained by the function and use of basic modal operators, which is to convey, enforce, or advocate for semantic rules or their consequences.[19] Thus, for normativism, our practice of introducing and eliminating basic modal operators is not explained by substantive modal facts, substantive possible worlds, or substantive states of affairs. As Thomasson argues at length, the normativist explanation is a more naturalistically acceptable account of what constrains the meaning of basic modal operators.[20]

In the next section, I will show that our introduction of singular terms for worlds and talk of similarity between them is not explained by their tracking substantive modal features of reality but by their normative function. That is, just like our use of basic modal operators, our literal talk of possible worlds is also fully explained by our use of that talk to convey, enforce, or advocate for semantic rules or their

[18] Thomasson (2024), Halliday (2009).

[19] Thomasson (2020: 81–82).

[20] See also Sidelle (1989) and Einheuser (2006).

consequences. It is in virtue of sharing these metasemantic grounds, that R1-R3 are conceptual truths. For this reason, I do not take literal talk of world to be more controversial than basic modal operators and so do not think that we must appeal to basic modal claims as pleonastic grounds for possible worlds.[21] Furthermore, an important consequence of this is that R1-R3 provide transformation rules for moving between literal talk about worlds and basic modal claims.

9.3.3 From Norms to Worlds, Ways, and Possibilities

So, how do we get from the normative function and use that the normativist argues underwrites our basic modal claims to talk about worlds, ways, or possibilities? Thomasson (2007b, 2013, 2020) develops modal normativism from the observation that the deontic use of modal operators provides a useful way of stating the rules of a game in an indicative rather than an imperative form.[22] For example, when communicating and reasoning about the rules of chess, the indicative modal claim "the pawn *must* never move backwards" is more useful than the imperative claim "do not move the pawn backwards!" because, for example, the former can more naturally be embedded in conditionals. Overall, a strong reason in favor of Thomasson's view of metaphysical modality is that it provides a unifying explanation of the function of deontic and alethic modal operators—to convey norms.[23] However, one might worry that it's difficult to see how talk about possible worlds (or ways or possibilities), which involves quantification and so appears to be descriptive, has any relevance to conveying norms.[24] So, why should the modal normativist care about literal claims about the existence of possible worlds? One reason is that, as I argued above, literal talk of ways, worlds, or possibilities nicely captures a distinction between metaphysical and epistemic modalities. In Sect. 9.4, I will argue that talk of distinct possibilities and talk of similarities between them is a useful way of conveying, mandating, or negotiating norms that can also be expressed using basic counterfactual claims. But first, I will show how literal talk of worlds, ways, and possibilities is explained by its regulative role in discourse by illustrating their use in games of *linguistic scorekeeping*.[25]

[21] This is one way my argument diverges from Thomasson (2020) and Steinberg (2013).

[22] Thomasson (2007b: 138).

[23] Thomasson (2013: 145).

[24] See Yablo (1996) for similar worries.

[25] Importantly, arguing that seemingly descriptive claims can also have a normative function in discourse is not *ad hoc*—see McGowan (2004).

9.3.4 *Talk of Worlds in Games of Linguistic Scorekeeping*

Drawing on Lewis (1983) and Brandom (2000), in a game of linguistic scorekeeping one's score is determined by those statements she is entitled or committed to. In any linguistic game, each speaker is keeping track of their own linguistic commitments and entitlements as well as the linguistic commitments and entitlements of other players. Overlap between players constitutes a *common score*. On the one hand, questions of *entitlement* are "upstream" from assertions or endorsements and come up when we are asked for reasons for making an assertion or endorsing a claim. So, e.g., if I am asked why I am entitled to claim, "The swatch is red," I can respond, "Because the swatch is crimson." On the other hand, *commitments* are "downstream" from assertions or endorsements. Brandom (2000) claims:

> ...for such a game...to be recognizable as involving assertions, it must be the case that playing one [statement], or otherwise adding it to one's score, can *commit* one to playing others, or adding them to one's score. If one asserts, "The swatch is red," one *ought* to add to one's score also "The swatch is colored." Making one move *obliges* one to be prepared to make the other as well (190–191).

A player's total score in this linguistic game is given by those statements that she is entitled to as well as those statements she is committed to. Basic modal claims have a number of important roles to play in games of linguistic scorekeeping. First, note that some of the statements in each player's total score will reflect what other statements are even up for potential play at all, i.e. what claims are assertible or endorsable, and these statements will often involve modal expressions. For example, suppose a claim of "possibly, *p*" is part of the common score. Then all of the players endorse "possibly, *p*" and even though none of them might yet endorse "*p*" it is an option allowed by the game. Importantly, this means that one can endorse *p* as a matter of fact or as a hypothetical supposition. Though, of course, once one attempts to endorse *p* as a matter of fact one is then expected to justify that endorsement in order to be granted the entitlement, i.e. before one can add *p* to one's scorecard. Alternatively, if a player manages to successfully add the basic modal claim "necessarily, *p*" to the common score, then not only will *p* become part of the common score but also *p* will be granted as an auxiliary supposition no matter what other suppositions are made.[26] This means that "not, *p*" is off limits and so cannot be played by any of the players, i.e. "not, *p*" may not be endorsed as a matter of fact or a standard hypothetical supposition.[27] Following McGowan (2004), this means that

[26] See also Thomasson (2020: 83–84).

[27] A hypothetical supposition that *p* is *standard* just in case *p* is possible in the context. A hypothetical supposition that *p* is *non-standard* just in case *p* is impossible in the context. In a given game non-standard suppositions are *sometimes* allowed depending on the context. These often come in the form of counterpossible conditionals, e.g. see Locke (2019).

in making a claim of necessity I am attempting to change facts about what is permissible in the particular conversational game we are playing.[28]

To see how this works, first consider an ordinary, practical modal claim rather than a robust metaphysical modal claim. Assume that the function of practical modal claims is to mandate or endorse a range of relevant actions. For example, in trying to coordinate a plan of action with others, I might assert, "I *must* leave work early if I am to make it to the bank on time and deposit my check." If accepted, it will mean that the statement "I will stay at work until 5 P.M. and deposit my check on time" is no longer up for play for any player, i.e. can no longer be proposed as a plan of action. So, one thing I am doing with my assertion is trying to change facts about what moves are allowed in our particular linguistic game, e.g. make it so that nobody will be permitted to endorse the claim that I will stay at work until 5 P.M. and deposit my check on time *either* as a matter of fact *or* hypothetical supposition. If I *must* leave work early if I am to deposit my check, then any contrary hypothetical supposition would be irrelevant given the context of our game.[29]

Suppose I make the move to endorse the claim of practical necessity because I believe it is not practically feasible for me to do both of those things. However, someone can challenge my move, by saying something to the effect of "possibly, not-*p*," e.g. they might say, "You *could* stay at work until 5 PM and still deposit your check on time." However, given the vagueness of this counter claim, I may demand that player say what entitles them to that claim by asking the following question: "But *how* could I do that?" At this point the player may reframe their original basic modal claim in quantificational terms by saying, "There are many *ways* you could do that—you could deposit your check online or you could have your partner come pick up the check and deposit it for you or you could hire a certified courier to deposit the check for you or you could live in world with teletransportation or you could live in a society such that full-service banks are open 24/7 or you could ...etc." So, the player has moved from using to basic modal claims in order to change facts about what moves are allowed in our particular linguistic game to introducing claims that quantify over and refer to *ways* the world could be in order to do the same thing.

Of course, I will not find just any of my interlocutors disjuncts as entitling her to the more general claim about what is possible or as a successful counter to my original move. It seems that not all of the distinct possibilities she raised were acceptable, likely because not all of the possibilities she is quantifying over are relevant given the (practical) context. If we press on why some of the possibilities offered in

[28] McGowan (2004) argues that some utterances are not only descriptive but change facts about what is permissible in a conversation. At this point nothing I've said is inconsistent with substantive view of modality. We often do two things at once with our utterances. For the modal normativist, not only are players using metaphysical modal claims to enforce or change facts about what statements can be played but they are using metaphysical modal claims to enforce semantic rules governing the term involved in those claims.

[29] See Lewis (1983) for more discussion of relative modalities in language games.

the example are relevant but not others, it seems that we will start talking about the ways in which the possibilities listed are more or less similar to actuality, e.g. "A society with teletransportation (much less 24/7 full-service banking!) would be nice, but our society is *nothing like that*!" In other words, we often make comparison claims about possibilities. This will be important in Sect. 9.4 when I talk about counterfactuals.

But setting this complication aside for now, suppose that I accept at least the first disjunct. If that is the case, then my original move is successfully countered and so I will not be able to add my original modal claim the common score of the game. Instead, I am justified in adding the claim "*There is a possibility* that I will stay at work until 5 PM and still deposit my check on time, say, by depositing my check online." Furthermore, this entitles me to the more general modal claim "I *can* stay at work until 5 and deposit my check on time."

In sum, in the linguistic game of making, accepting, and rejecting assertions, basic modal claims play a role in limning what claims are up for play to begin with, i.e. what claims may be endorsed as a matter of fact or hypothetical supposition. But I have just shown that literal talk of ways (worlds or possibilities), which on their grammatical surface are descriptive and not normative, play the same role of enabling speakers to change permissibility facts in a game of linguistic scorekeeping. Attempts by a player to add the claim "necessarily, *p*" to the linguistic score is an attempt to change permissibility facts, namely that "not, *p*" may no longer be played by any of the players. Alternatively, if "*not*, p" may not be endorsed as a matter of fact or hypothetical supposition, then players can just as well add the universal affirmative "for any way things might go, *p*" to the common score. Either way, these attempts can be challenged by an opponent with existential claims about distinct possibilities or ways the world could be. If the opponent is successful, then "there is a possibility that not-*p*" is added to the common score which means that any player will be allowed to try and endorse "not-*p*" as a matter of fact or hypothetical supposition.

So, players can introduce talk of distinct possibilities in order to regulate a game of linguistic scorekeeping and to counter attempts to remove certain claims from play. A substantive explanation is that players are able to introduce talk of possibilities because that talk tracks substantive modal properties or worlds. But to require this would grossly mischaracterize this use of world talk. If someone were to interject on the game by demanding players explain where all of these so-called "ways" and "possibilities" are, they would be met with confused looks in much that same way someone would be met if they claimed that they understood that I bathed but do not yet understand what happened to the bath after I took it. Instead, a simpler explanation is that players are able to use talk of distinct possibilities because it allows them to regulate the game of linguistic scorekeeping and to counter attempts to remove certain claims from play and because that talk is merely a syntactic variation of basic modal claims, which are also better explained by their regulative role.

9.3.5 Talk of Worlds in Mandating Semantic Rules

The example I just gave involved a practical sense of modality. Claims of metaphysical modality play a similar role in games of linguistic scorekeeping, but instead of regulating what courses of action are relevant, metaphysical modality regulates semantic rules. Again, attempts to add the claim "necessarily, *p*" to a linguistic score are attempts to mandate permissibility facts within the game, namely that "not, *p*" may not be played in the game, i.e. endorsed as a matter of fact or hypothetical supposition. But, for the normativist, the case of metaphysical modality is especially important because modal claims *also* mandate semantic rules and their consequences, i.e. facts how we are allowed to apply the expressions at any point of the game. When someone successfully adds "necessarily, bachelors are unmarried" to the common score of the game, "bachelors are unmarried" will become part of the common score and may be endorsed as a hypothetical supposition no matter what other suppositions are made. Furthermore, this means that any claim that entails "there is a married bachelor" will be off limits and so cannot be played by any of the players. So, suppose that "Kim is married" is part of the common score, then "Kim is a bachelor" will be off limits and also cannot be endorsed as a hypothetical supposition so long as "Kim is married" is held fixed. Importantly, adding the metaphysical necessity to the common score mandates how the term "bachelor" can be applied in further moves made in the game.

Consider the *a posteriori* necessity "necessarily, water is H_2O." For the normativist, this conveys two things, one conceptual and the other empirical. The *conceptual part* is that the application conditions for "water" are *defined* in terms of facts about its actual chemical composition. The *empirical part* is the scientifically discovered fact that the actual chemical composition of the watery stuff in our environment is H_2O. Combined, we get the basic modal claim of necessity about water. When chemists add "necessarily, water is H_2O"—or its equivalent "it is not possible that water is not H_2O"—to the linguistic scorecard of chemistry this reflects that "water is not H_2O" is off limits, i.e. not up for play in the game. Further, by adding the modal claim, facts about what suppositions or hypotheses are permissible in the game change. This means that no matter what other claims are added or taken away from the linguistic game "water is not H_2O" can never be added to the game because it violates some consequence of the semantic rules that govern the use of "water" with respect to actual conceptual or chemical facts. Importantly, this also forecloses further conceptual inquiry about the meaning of "water" (e.g. is it a chemical kind?) as well as empirical suppositions about the watery stuff (e.g. what is the chemical composition of water on Twin Earth?).

But, again, all of this can be just as well conveyed with syntactic metaphors using talk of worlds. The semantic rules and their consequences are just as well be conveyed in the object language using the relevant terms along with introducing literal talk of ways the world can or cannot be. One way is with nominal talk of worlds in the form of universal affirmatives such as "for any way things might go, *p*". For example, introducing "For any way the world might be, water is H_2O" to the common score mandates certain permissibility facts about hypothetical suppositions, namely that "water is H_2O" will be granted as an auxiliary supposition no matter what other suppositions are made because otherwise the rules that govern the use of "water" with respect to the actual conceptual or chemical facts will be violated. Alternatively, we might introduce the negative existential "There is no way that water could not be H_2O—there is no way that water could be XYZ, there is no way that water could be YZX, there is no way that water could be a mere idea in the mind of the Author of Nature, etc." Further, like the basic modal claim, this denial of the existence of ways the world might be forecloses any conceptual or empirical inquiry into any one of these individual claims about a way water could be something other than H_2O. For example, one is not permitted to try and play the claim "there is a possibility such that water is a mere idea in the mind of the Author of Nature" while still counting as a player in the particular game the chemists are playing—such a claim would violate a semantic mandate on the use of the term "water". However, insofar as one insists, then one can be seen as trying to negotiate changes to the rules of the game, i.e. changes to the semantic rules governing the term "water".[30]

So, for modal normativism, to universally quantify over metaphysical possibilities or to deny the existence of a certain metaphysical possibility is to mandate what is permissible when applying our terms to actual or hypothetical scenarios. To literally say, "there is no possibility or way the world could be such that a flattened statue survives" just is to mandate the rules that govern the use of the term "statue". This is the exact same semantic explanation for what it is to literally say, "it is not possible for a flattened statue to survive". So, basic modal claims and literal talk of worlds share the same exact metasemantic normativist grounds, and the only difference between these claims is syntactic. This is why R1-R3 are conceptual truths for the modal normativist.

9.4 Individuating Possibilities and Talk of Similarity for Normativists

The syntactic difference between basic modal claims and literal talk of worlds provides us with different linguistic forms with which to mandate and convey semantic rules. For example, suppose someone asserts that "possibly, Tabby the cat is on a mat" which conveys both that "Tabby is on a mat" is up for grabs as a hypothetical supposition as well as conveys permission about the use of "Tabby".

Technically, Tabby could be on a mat in the living room, Tabby could be on a mat in the bathroom, Tabby could be on a mat in the Oval Office in the White House, Tabby could be on a mat in a space station on Mars, Tabby could be on a flying mat, Tabby could be on a blue mat, a red mat, etc. So, that's at least *six* different possible

---

[30] Thomasson (2020: 204–205), Locke (2020).

ways Tabby could be on a mat (and maybe one impossible way). Furthermore, a literal assertion of any of these distinct ways the world might be is fully explained as an expression of a potential permission of the application of "Tabby" but while remaining in the object language using the term rather than mentioning it.

However, in addition to making literal claims about possibilities, we often make comparison claims about those possibilities as well. We can ask which of a number of considered possibilities are more or less similar to how things actually are. For example, perhaps the possibility of Tabby being on a mat in the living room is more similar to how things actually are than the possibility of Tabby being on a mat in the Oval Office of the White House because Tabby is actually on a sofa in the living room of a house in Canberra, Australia. Below I will argue that that these similarity claims also reflect permissions and rules of use and so are fully explained by their normative role in expressing important features of semantic rules.

Since we make literal comparison claims about possibilities, this means we have some idea of how these possibilities are individuated in a context. But how does that work? I think a common mistake made by metaphysicians is losing sleep on how to individuate worlds in terms of the nature of these worlds, e.g. as states of affairs, mereological sums, as sets of sentences, as pictures, propositions, etc. But the theoretical pressure to individuate worlds in these terms, and so provide context independent identity conditions for these entities, comes from positing worlds as theoretical entities needed to explain what makes modal claims true, or what the content of beliefs amounts to, or what the content of propositions amounts to, etc. But for the normativist, talk of ways, worlds, and possibilities functions to convey, mandate, or negotiate semantic rules and that's it.[31] Thus, these entities need not have any existence beyond their instrumental use in this practice. This means that at most, ways, worlds, and possibilities are syntactic artifacts in a (mathematical) apparatus used to index claims about what is possible, which on the normativist view amounts to no more than indexing features of semantic rules of use such as in what contexts permissions will be granted for the use of a term or when the rules will be strictly enforced no matter what. Nonetheless, I will argue that in practice these ways, worlds, or possibilities are individuated by their correspondence with sets of auxiliary assumptions relevant in the context, e.g. when evaluating counterfactual claims.

Judgements comparing distinct possibilities mirror counterfactual judgments about just how different the way things actually are *would* have to be, just how much the way things actually are *would* need to change, in order for the different possibilities to count as actual. I will argue that, for modal normativism, these counterfactual judgments are explained by a particular kind of semantic competency that involves being able to consistently apply non-modal terms, according to the semantic rules that govern those terms, across a range of sets of auxiliary assumptions. I will then

[31] Of course, someone might fault the normativist for this and argue that the normativist fails to take advantage of all the theoretical usefulness of possible worlds (discussed in Lewis, 1986). My goal in this paper is not to defend normativism from these kinds of objections but to show how normativism can keep literal talk of worlds.

show how literal talk about the possibility that *p* being distinct from the possibility that *q*, in a given context, is both merely a syntactic reflection of this semantic competency over different sets of auxiliary assumptions and the ability to judge some sets of auxiliary assumptions as relevant to the application of a term but not others. The best way to illustrate this is by first showing how counterfactuals work for the modal normativist.

Recall that according to normativism, "Necessarily, *p*" serves to mandate, correct, or condemn uses of the terms involved in *p*. Given a normativist account of basic modal expressions, it is natural to think that the modal expression "would" plays a correlative *normative* role of mandating, correcting, or condemning uses of the terms in conditional claims of the form "if *p* were the case, then *q would* be the case".[32] Claims such as "Possibly, *p*" convey that the terms in *p* are used in a permissible way. Thus, given a normativist account of basic modal expressions, it is natural to think that the modal expression "might" plays a correlative *normative* role of expressing permissions in conditional claims of the form "if *p* were the case, then *q might* be the case". Thus, "would" and "might" serve to convey, mandate, or negotiate (but never to describe), semantic rules and permissions, or their consequences, of terms as they are used under hypothetical suppositions. This gives us the following account of would- counterfactuals for modal normativists:

N3. Statements of "if *p* were the case, then *q would* be the case" express a *semantic requirement* to accept that *q* given *p* as a hypothetical supposition and a range of *relevant auxiliary assumptions* determined by the context, e.g. statements of "if Kim were a bachelor, then Kim would be unmarried."[33]

Thus, the normative role of metaphysical counterfactuals has to do with conveying how *p* and *q* are properly used, i.e. with the semantic rules and permissions governing uses of *p* and *q*.

We might ask: when does "if *p* were the case, then *q would* be the case" fail to express a semantic requirement? On a certain strict conditional reading of counterfactuals, the answer is: *never*. Earlier expressivist work on counterfactuals treated them as "inference tickets" and focused on the lawlike nature of certain counterfactual claims. In particular, for these early expressivists, lawlike dispositional statements or lawlike counterfactuals express *season* inference tickets, i.e. inference tickets that are good any time of year so to speak.[34] The less metaphorical way of making this point is that a "season" inference between the antecedent, *p*, and the consequent, *q*,—an inference which for normativists partly constitutes the meaning

[32] Indeed, for a long time a common, if not standard, analysis of *would*-counterfactuals was as strict conditionals of the form $\Box(p \supset q)$.

[33] Importantly, this is for metaphysical counterfactuals and not epistemic counterfactuals (see Vetter, 2016 for more on this distinction). Thus, for the normativist the requirement is *semantic* insofar as one *must* use the terms as they are used in *q* given *p* as a hypothetical supposition. It is not an *epistemic* requirement to believe *q* given a belief in *p*.

[34] See Ryle (1950, 2000), Sellars (1958).

of p and q—is an inference that is *maximally* robust insofar as the addition of *any* range of relevant auxiliary hypotheses will not infirm the inference between p and q.

For example, in the context of nomological counterfactuals, one way of reading Sellars (1958) is as claiming that the only purpose of introducing talk of causal necessity into our discourse, as we might with '$\Box_C (p \rightarrow q)$', is to express that the inference from p to q is a meaning constituting inference that holds no matter what other assumptions are made. This idea is also reflected in the modal normativist view of modal operators of metaphysical necessity insofar as '$\Box(p \rightarrow q)$' can be taken as a way of illustrating, endorsing, or expressing that $p \rightarrow q$ reflects some semantic rule, or consequence of some rule, that governs the use of p and q, e.g. "If someone is a bachelor, then that person is unmarried." Again, this is because the inference from p to q is a meaning constituting inference that holds no matter what other assumptions are made. But, when talking about the normative force of "would" in *counterfactuals more generally*, we need to be careful, because a counterfactual claim $p \Box\!\rightarrow q$ is not necessarily equivalent to a strict conditional $\Box(p \rightarrow q)$. For the normativist, the latter may indeed express a sort of *everything*-considered counterfactual robustness of some meaning constituting relation between the antecedent and the consequent, i.e., a connection between p and q with "full-stop" normative force. But not every useful or interesting metaphysical counterfactual (i.e. a counterfactual used to express semantic rules) needs to be *everything*-considered robust. Very often contextually limited robustness is also normatively relevant. As I will argue below, having a grasp on even these weaker counterfactuals, i.e. an understanding of even *some*-things-considered robustness, can be equally important for counting as a competent user of ordinary empirical terms such as "cat".

So, we need to accept that the requirements expressed with counterfactual claims often depend on *contextual* considerations—a requirement expressed by a counterfactual may not hold "full-stop" insofar as it may hold in one context but not hold in others. For example, suppose that in some context "if the match were struck, it would ignite" successfully expresses a requirement to accept that a match ignites on condition of the acceptance that the match is struck. In that case, suppositions such as the matches are dry, that the matches are not struck in a vacuum, etc. are held fixed. However, in different contexts, the requirement expressed by the counterfactual might not hold. This is because there are contexts involving different auxiliary assumptions—for example, that the match is underwater, that we are in a vacuum, etc.—that can easily undermine the requirement expressed in the first context. There are at least two ways this undermining can happen. It can happen *explicitly*, as when I say, "If the match were struck *in a vacuum*, it would *not* ignite," but that need not be the case. It could be that we are in a context where we are already talking about vacuums, say a context where we are talking about space walking astronauts. In this case, "if the match were struck, it would ignite" fails to express a requirement to accept "the match ignites" given the hypothetical acceptance of "the match is struck" because we are making auxiliary hypothetical assumptions that infirm the inference from antecedent to the consequent.

Of course, it could be that certain auxiliary hypothetical assumptions are *irrelevant* to the evaluation of a counterfactual in a given context. For example, suppose that while explaining to a hiker that campfires are not permitted in the forest during a long-lasting drought, a ranger says, "if the winds were to blow embers from the fire, part of the forest would catch fire and burn down". Then it would not be appropriate for the hiker to counter by saying, "But there is a possibility that a sudden rainstorm would put the fire out." Here the hiker is trying to introduce an auxiliary assumption into the context about rainstorms. However, since the ranger is making the claim during a long-lasting drought, the hiker's response is inappropriate because it is trying to introduce an auxiliary assumption that is *irrelevant* to the ranger's claim. Thus, the range of *relevant auxiliary assumptions* are those assumptions we hold fixed and are needed to accommodate a hypothetical supposition that *p* without deviating *too far* from what is the case in the circumstance of utterance.[35]

So, for the normativist, metaphysical counterfactual judgments about what *would* happen *if* mandate semantic rules and are based on our judgments about semantic requirements to accept certain claims given (i) a hypothetical supposition and (ii) various ranges of auxiliary assumptions determined by the context. These judgements reflect semantic competency with ordinary non-modal vocabulary and requires that speakers and hearers be able to determine both which sets of auxiliary assumptions are relevant and for which sets the semantic rules apply. For example, the counterfactual claim "if the clay statue were flattened, then the statue would be destroyed but not the clay" mandates an *everything*-considered rule that applies across all sets of relevant auxiliary assumptions.

Importantly, we can mandate this semantic rule just as well using literal talk of worlds. For example, a statement of "for every *way* that you might flatten a clay statue, the statue will be destroyed but not the clay—it will be destroyed if you flatten it on Tuesday but not Monday, if you flatten it on Mars but not on Earth, if you flatten it in a black hole but not in a machine press, etc." mandates that for any relevant set of auxiliary assumptions that are adopted, in addition to the supposition that a clay statue is flattened, that set *must* include that the statue is destroyed but not the clay. Of course, the sets must be *relevant* and semantic competency requires that speakers and hearers deem certain assumptions irrelevant. For example, an auxiliary assumption that to be a statue is to be a certain immaterial substance that persists through any physical change is irrelevant given a semantic framework where the application conditions of the sortal "statue" are defined in terms configurations of ordinary physical materials and not immaterial substances. Again, this auxiliary assumption can be ruled out using literal talk of ways, worlds, or possibilities—"there is no possibility where statues are an immaterial substance that persist through any physical changes"—or even with talk of impossibilities—"there is a world with a statue that is an immaterial substance that persists through any physical change, but that world is impossible and nothing like the actual world." According to the normativist account of metaphysical counterfactuals I just provided, counterfactual claims are fully explained by (a) the practice of mandating semantic rules across sets of auxiliary assumptions and (b) the ability to determine which sets are

[35] See Stalnaker (1968: 104) and Lewis (1973: 8–9). Often determining relevant auxiliary hypothetical assumptions is a matter of trading off between similarity and differences with the circumstance relevant for evaluating a counterfactual.

relevant. Since talk of distinct possibilities and talk of similarity between them are merely syntactic variations of those counterfactual claims, talk of distinct possibilities and talk of similarity between them is also fully explained the same practice and ability. In practice, when we individuate between possibilities we are individuating between different sets of auxiliary assumptions.

Importantly, semantic competency is not just reflected by *everything*-considered judgements, i.e. the judgment that the semantic rule applies no matter what set of relevant auxiliary assumptions we consider. Semantic competency also requires the more fine-grained skill of ranking a range of sets of assumptions in terms of relevancy in the context. For example, consider the name of a cat "Tabby" and the required judgement that if Tabby were on a mat, then Tabby would be quite happy but not so if that mat were on the surface of the Sun. Brandom (2008) explains these kinds of judgments in terms of what he calls the Kant-Sellars thesis:

> One has not grasped the concept <u>cat</u> unless one knows that it would still be possible for the cat to be on the mat if the lighting had been *slightly different*, but not if all life on earth had been extinguished by an asteroid-strike (97).

Of course, our judgments given less extreme auxiliary assumptions are just as important reflections of semantic competency. For example, assuming the way things actually are, with Tabby comfortably on a couch in Canberra, Australia, we should reject the counterfactual, "if Tabby were on the mat, then Tabby would be in the Oval Office." This is because things would need to change considerably more in order for Tabby to be on a mat in the Oval Office in the White House than for Tabby to be on a mat in the living room. In other words, in addition to making the hypothetical assumption that Tabby is on the mat, we would have to accept auxiliary assumptions that are simply not relevant.

Again, literal talk of similarity between distinct possibilities is merely a syntactic reflection of this kind of linguistic competency with the term "Tabby" and corresponding empirical claims such as "Tabby is on the couch." Suppose we are in Canberra, Australia. If we endorse the empirical claim "Tabby is on the couch" we are thereby committed to making a semantic distinction between the set of assumptions (a), which includes the supposition that Tabby is on the mat and Tabby is in the living room, and another set of assumptions (b), which also includes the supposition that Tabby is on the mat but that Tabby is in the Oval Office in the White House. The semantic distinction between (a) and (b) is underwritten by two semantic judgments: (i) that (b) reflects a gross deviation from how we ordinarily apply the term "Tabby" and (ii) that (a) is relevant to the application of "Tabby" but not (b). All of this is can be expressed in the object language with the comparative claim "The possibility that Tabby is on a mat in the living room is more relevantly similar to how things actually are than the possibility of Tabby being on a mat in the Oval Office in the White House." Presumably, someone who rejects this claim would have strange beliefs about the world underwritten by either empirically false beliefs about cats or a deviant understanding of the proper use of the term "cat," e.g., maybe they think that cats are *sorts* of creatures intrinsically capable of teleportation.[36] In that case,

[36] In a metalanguage they associate the term "cat" with the sortal term "things that teleport."

we have good reason to suspect that person is not competently using the term "Tabby" and is not entitled to the claim "Tabby is on the couch".

So, talk of worlds and similarity is merely a syntactic metaphor of counterfactual claims. For normativism would-counterfactuals are fully explained by a semantic requirement to accept the consequent given the antecedent as a hypothetical supposition and a range of relevant auxiliary assumptions determined by the context. Furthermore, semantic competency requires being able to individuate different sets of auxiliary assumptions and assess their relevance. This is all that talk of worlds and similarity amounts to since such talk is merely a syntactic metaphor of counterfactual claims. Thus, talk of worlds and similarity is fully explained by the same normative function—to convey, mandate, or negotiate semantic rules of use.

9.5 Conclusion

In this paper I have argued that modal normativism can and ought to keep literal talk of possible worlds and similarity between them. On my view, talk of worlds and similarity is merely a syntactic device for conveying, mandating, or negotiating semantic rules. Just as this function fully explains introducing basic modal operators, I have argued it fully explains literal talk of ways, worlds, and possibilities. I also argued that ways, worlds, and possibilities are individuated by their correspondence with sets of auxiliary assumptions relevant in the context.

Furthermore, by characterizing talk of ways, worlds, and possibilities as primarily a syntactic tool indexing and formalizing basic modal claims normativism allows us to keep literal talk of possible worlds without the awkward task of fitting substantive worlds or states of affairs into a naturalistic ontology.

Finally, since evaluating basic modal claims comes down to conceptual analysis and straightforward empirical inquiry, evaluating claims about ways, worlds, and possibilities comes down to the same thing. Hence, we have a plausible epistemic story for how we know claims about worlds, e.g. I know there is a possibility where I exist but not my biological parents because of the semantic rules that govern the relevant terms.

References

Blackburn, S. (1993). Morals and modals. In *Essays in quasi-realism*. Oxford University Press.

Brandom, R. (2000). *Articulating reasons: An introduction to inferentialism*. Harvard University Press.

Brandom, R. (2008). *Between saying and doing: Towards an analytic pragmatism*. Oxford University Press.

Einheuser, I. (2006). Counterconventional conditionals. *Philosophical Studies, 127*(3), 459–482.

Forbes, G. (1985). *The metaphysics of modality*. OUP.

Hale, B., & Wright, C. (2001). *The reason's proper study: Essays towards a Neo-Fregean philosophy of mathematics*. OUP.

Halliday, M. (2009). The essential halliday, ed. Jonathan J Webster, Continuum.
Lewis, D. K. (1973). *Counterfactuals*. Blackwell.
Lewis, D. K. (1983). Scorekeeping in a language game. In *Philosophical Papers* (Vol. I). Oxford University Press.
Lewis, D. K. (1986). *On the plurality of worlds*. Oxford University Press.
Liggins, D. (2014). Grounding and the indispensability argument. In D. Molinini, F. Pataut, & A. Sereni (Eds.), *Synthese*, special issue Indispensability and Explanation. https://doi.org/10.1007/s11229-014-0478-2
Locke, T. D. (2018). *Counterpossibles for modal normativists*. Ph.D. dissertation, University of Miami.
Locke, T. D. (2019). Counterpossibles for modal normativists. *Synthese*. https://doi.org/10.1007/s11229-019-02103-1
Locke, T. D. (2023). Thomasson on ontology. In M. Garcia-Godinez (Ed.), Modal normativism and metasemantics (pp. 109–136). Springer International Publishing Cham.
Locke, T. (2020). Metaphysical explanations for modal normativists. *Metaphysics, 3*(1), 33–54. https://doi.org/10.5334/met.35
McGowan, M. K. (2004). Conversational exercitives: Something else we do with our words. *Linguistics and Philosophy, 27*(1), 93–111.
Plantinga, A. (1974). *The nature of necessity*. Oxford University.
Press Rosen, G. (1990). Modal fictionalism. *Mind, 99*, 327–354.
Ryle, G. (1950). 'If,' 'so,' and 'because'. In M. Black (Ed.), *Philosophical analysis, A collection of essays* (pp. 323–340). Cornell University Press.
Ryle, G. (2000). *The concept of mind*. University of Chicago Press.
Salmon, N. (1988). Review of on the plurality of worlds. *Philosophical Review, 97*(2), 237–244.
Sellars, W. (1958). Counterfactuals, dispositions and the causal modalities. In H. Feigl, M. Scriven, & G. Maxwell (Eds.), *Minnesota studies in philosophy of science, Volume 2: Concepts, theories and the mind-body problem* (pp. 225–308). University of Minnesota Press.
Sidelle, A. (1989). *Necessity, essence, and individuation: A defense of conventionalism*. Cornell University Press.
Stalnaker, R. (1968). A theory of conditionals. *American Philosophical Quarterly*, 98–112.
Steinberg, A. (2013). Pleonastic possible worlds. Philosophical Studies, 164(3), 767–789. https://doi.org/10.1007/s11098-012-9857-z
Thomasson, A. (2007a). *Ordinary objects*. OUP.
Thomasson, A. (2007b). Modal normativism and the methods of metaphysics. *Philosophical Topics, 35*(1-2), 135–160.
Thomasson, A. (2009). Answerable and unanswerable questions. In D. Chalmers, D. Manley, & R. Wasserman (Eds.), *Metametaphysics: New essays on the foundations of ontology* (pp. 444–471). OUP.
Thomasson, A. (2013). 2012 Nancy D. Simco lecture: Norms and necessity. *Southern Journal of Philosophy, 51*(2), 143–160.
Thomasson, A. (2014). *Ontology made easy*. OUP.
Thomasson, A. (2020). *Norms and necessity*. OUP.
Thomasson, A. L. (2024). How should we think about linguistic function? Inquiry (Oslo), 67(3), 840–871. https://doi.org/10.1080/0020174X.2022.2074886
Vetter, B. (2016). Counterpossibles (not only) for dispositionalists. *Philosophical Studies, 173*, 2681–2700.
Warren, J. (2016). Sider and the epistemology of structure. *Philosophical Studies, 173*(9), 2417–2435.
Warren, J. (2017). Epistemology versus noncausal realism. *Synthese, 194*, 1643–1662.
Yablo, S. (1996). How in the world? *Philosophical Topics, 24*(1), 255–286.

Chapter 10
A Pragmatic Approach to (Some) Disputes in Semantics

Irene Olivero

Abstract Quite recently, several ongoing disputes in semantics have gained attention. Some of these disputes include questions about whether we would apply the term 'pencil' to objects that satisfy the superficial description of the properties associated with classic artifactual pencils but are in fact a species of organisms; whether 'woman' applies only to females; or whether the rules governing the application of terms for natural substances and species (e.g., 'water', 'tiger', gold', etc.) are the same as those governing the application of words for artifacts and artifactual kinds (like 'chair', 'pencil', 'television', etc.) or social and institutional kinds (e.g., 'pediatrician', 'university', 'parent', etc.). These disputes in semantics, I argue, are also disputes about what the referents in question are--that is, they are also metaphysical disputes. The aim of this paper is to show how abandoning the traditional "heavyweight meta-metaphysical realism" within which such disputes are typically framed, in favor of Amie L. Thomasson's normativist, pragmatic approach, allows us to see them as "metalinguistic negotiations". That is, these are best understood as disputes about how the words in question *ought* to be applied rather than about how they are in fact applied. I thus highlight several advantages of this pragmatic perspective, which I label a "Negotiating Approach". It offers a more viable and appealing explanation of why such disputes arise in the first place, it avoids the epistemological mysteries faced by traditional accounts, and provides a more promising attempt to address or (dis)solve such quarrels. Moreover, I argue that this approach reveals that meaning is a variable whose value can be negotiated according to pragmatic and empirical factors (including nature, authority, ethics, and other contextual considerations.). This recognition is crucial for its further pragmatic import, which represents an additional advantage of the approach proposed: meaning can be revised and engineered in line with these relevant factors.

I. Olivero (✉)
DIG Department, Polytechnic University of Milan, Milan, Italy
e-mail: i.olivero@gmail.com

M. Sambrotta (ed.), *Metaphysics Today: In Conversation with Amie Thomasson*, Synthese Library 512, https://doi.org/10.1007/978-3-031-98468-6_10

Keywords Semantic disputes · Metaphysical disputes · Artifactual kind terms · Pragmatic approach · Metalinguistic negotiations

10.1 (Some) Disputes in Semantics

A recently debated issue in semantics concerns whether Hilary Putnam's celebrated theory of direct reference (cf. 1973, 1975), originally developed for terms referring to natural substances and species (such as 'water', 'tiger', gold', etc.), can also account for the semantics of terms for artifacts and artifactual kinds (e.g., 'pencil', 'table', 'television', etc.). While it is widely accepted that natural kind terms are directly referential expressions—meaning that their extension is determined by the *real nature* of their referents, regardless of whether we know or will ever know what that nature is—it is much more controversial that this holds for artifactual kind term.[1] [new paragraph] Putnam managed to convince many philosophers of his thesis about natural kind terms by offering the following, seemingly quite persuasive picture[2]:

Dispute 0. 'Water'/Twin-'water'. What does the word 'water' apply to? Putnam asks us to imagine what we would say if we were to discover (say, on a different planet—the planet Twin Earth) a liquid with the same superficial characteristics listed in the description associated with the term 'water' but that has a different microstructure (it's not H_2O). Putnam remarks that we would not call it 'water'. This is because—contrary to what the traditional account of reference (also known as "descriptivism"[3]) holds—'water' (and natural kind terms in general) does not apply to whatever matches the description of the properties typically connected to the term (e.g., "colorless, odorless, tasteless liquid that quenches our thirst"). 'Water' rather applies to whatever shares a common *nature* with what we have paradigmatically indicated by that term (that is, H_2O). This point is further attested

[1] Putnam's direct referential theory has become probably the most shared image of reference concerning terms for natural substances and species (such as 'water', 'tiger', 'gold', etc.). This theory holds that there is a non-mediated connection between a natural kind term and its referent, and such direct, objective, external relation determines the term's reference—independent of whether we know or could possibly ever know what this relation actually is. Such theory appeals to (what is usually regarded as) the following strong intuition: (I) something (a natural kind) is correctly called 'P' if and only if it shares the *same nature* (essence, deep hidden structure, etc.) of most of the things that are paradigmatically called 'P', *not* if it matches the description typically associated with it. Moreover, this is so even though such a nature can be utterly mistaken by or completely unknown to all speakers.

[2] Consider a contemporary case that may seem very convincing. '*Covid-19*'. What does 'Covid-19' apply to? Currently, we probably still know only some properties that characterize this virus. Nonetheless, when we use that word, we refer to whatever shares the same nature as what was originally identified as Covid-19, even though we do not know exactly what this nature is, and despite the fact that new properties are being discovered with time.

[3] The main exponents of this traditional view notoriously are G. Frege (1892) and B. Russell (1905).

by the following fact. There was a time (before the discovery of chemistry) when the two liquids were indistinguishable from one another. Yet—Putnam insists—we would not use the term 'water' to refer to both the H_2O-liquid and the twin-"water", simply because back then we did not have epistemological access to what their respective natures (i.e., their chemical structures) were. 'Water' (and similar terms) is *rigid* and *indexical*: *rigid*, for it refers to the same objects across all possible world in which it refers (if it refers at all), and *indexical* because it designates whatever has the same nature as the paradigmatic samples (cf. Putnam, 1973, 1975).[4]

As noted, Putnam's theory initially seemed convincing to many. But in the same influential article (1975), he extended his semantic picture to other terms. This move raised the question: Is this picture equally persuasive in explaining the semantics of, e.g., artifactual kind terms? Disagreement on the answer has sparked an ongoing debate that revolves around a series of emblematic examples and imaginary cases.[5] The following are examples of typical disputes in this respect:

Dispute 1. 'Pencil'/Pencil-organisms. What does the word 'pencil' apply to? Does it have the same semantics as 'water'? Putnam (1975) answers this question affermatively. To support his claim, he asks us to imagine what we would say if we found out that, on a different planet, all the things that *pass* for pencils (for they share the same superficial characteristics with our pencils) are actually organisms. He claims that we would say: "The things that on that planet pass for pencils are not really pencils. They *are* a species of organisms". By contrast, if we were to discover that *all* pencils on Earth turned out to be organisms we would say: "*Pencils* are organisms". This is because, in that case, we would have discovered the real nature of *our* pencils to be biological rather than artifactual. This—Putnam explains—shows that artifactual kind terms behave like natural kind terms "with all the concomitant rigidity and indexicality" (1975, 244). What governs the reference of these terms is the "nature" of the actual objects they designate; it is not the properties listed in the description connected to these words. By contrast, Stephen Schwartz (1983) maintains that, in both cases, we would instead say: "Some pencils are

[4] Evidently, Putnam's notion of "indexicality" is a peculiar one, distinct from the common notion of indexicality tight to David Kaplan (1989). In Putnam's theory, an indexical term is one that refers to things that share the *same nature* as those objects that serve as paradigms (in the *actual* world). The notion of indexicality is then defined also in terms of rigidity: once the relevant object is picked out in the actual world, the reference of a given term no longer shifts from possible world to possible world. I have explored this peculiarity in more detail in another paper.

[5] The main participants of this debate include: S. Schwartz (1978, 1980, 1983); H. Kornblith (1980, 2007); J. Nelson (1982); D. Putman (1982); B. Abbott (1989); C. Elder (1989, 2007); A. L. Thomasson (2003, 2007a), D. Marconi (2013). A response to this debate can be found in I. Olivero (2018, 2019); Olivero and Carrara (2021). Olivero's (2018) essay proves that Putnam's argument—like the other main directly referentialist arguments advanced in this debate—ultimately fails in defending that artifactual kind terms have an externalist semantics. In turn, Olivero's (2019) article shows that artifactual as well as social or institutional kind terms cannot in fact have direct reference.

organisms". For, in the second scenario, nothing would prevent us from manufacturing artifactual pencils (cf. Schwartz, 1978, 1980, 1983).[6]

Dispute 2. 'Pencil'/Pencil-shaped devices for control. To defend Putnam's view, James Nelson (1982) imagined the following scenario: what would we say if someday we were to discover that we have been fooled by malevolent aliens who made us believe that pencils are what we think they are, i.e., tools designed for writing, marking, and drawing, whereas they are in fact devices by means of which those aliens manipulate our activity? We would answer—Nelson claims—: "The things [...] that pass for pencils [*are not pencils*; they] are really devices of control and manipulation" (1982, 362). The reference of 'pencil' and similar words is governed by the "nature" of the objects that we denote by these words, whether or not we know or will ever know what this "nature" is. By contrast, Schwartz insists that this case has nothing in common with the case of 'water' and other natural kind terms. He insists that, in the case of artifactual kind terms, the original specification "dominates" possible "empirical discoveries of the nature of the previous members in determining whether new candidates are members of the extension of the term" (1983, 478). In other words, Schwartz emphasizes that we continue to apply these terms to objects matching their original descriptions, regardless of later empirical discoveries about the paradigmatic examples of their designated kinds.

Consider now similar semantic disputes about social kind terms (such as 'pediatrician', 'university', 'money', etc.):

Dispute 3. 'Pediatrician'/Martian spies-pediatricians. What does the term 'pediatrician' apply to? Does it have the same semantics as 'water'? Putnam once again suggests that this is the case. In fact, this suggestion has given rise to a debate similar to the one about artifactual kind terms. Putnam and Nelson on one side, and Schwartz on the other, have discussed what our reaction would be if someday we discovered that what we thought were pediatricians were actually Martian spies. Putnam and Nelson think that in such a situation, we would describe this hypothetical scenario as one in which we had discovered the real "nature" of pediatricians. We would consequently abandon the link between the word and the definition associated with it, which, in their view, shows that also social kind terms refers to whatever shares the same "nature" as their paradigmatic referents. By contrast, Schwartz claims that we would describe this fictional case as one in which we had merely found additional skills (or degrees) of pediatricians. The fact that pediatricians are actually humans is something that may be somehow implicit in the definition, but it is not taken to be an *essential* requisite.

At this point, one may think that the disputes about artifactual and social kind terms can simply be dismissed, for it is not convincing that these terms have a directly referential semantics. Given the objections raised to this hypothesis, it seems reasonable to agree that these words apply to whatever matches the

[6] It can be noted that, in fact, the scenario described by Putnam is not one in which we have uncovered the real "nature" of pencils, but rather one in which we have discovered that 'pencil' is not an artifactual but a natural kind term, since it applies to a natural kind (cf. Olivero, 2018, § 3).

descriptions typically associated with them.[7] This dismissal still supports retaining a directly referential picture to explain the semantics of natural kind terms.

However, consider the following dispute:

Dispute 4. 'Cat'/'Key sparrow'. Putnam (1970, 1975) holds that, if we were to discover that *all* the cats on Earth had always been robots (remotely controlled from Mars), we would claim: "*Cats* have turned out to *be* robots". Accordingly, we would continue to apply the term 'cat' to all the objects to which we have always applied it, provided that they share the same "nature" as the paradigmatic robotic cats. This would again suggest that terms for natural substances and species are directly referential expressions.[8] By contrast, Amie L. Thomasson argues that this is not the case. She does so with her 'Key sparrow' hypothetical case. "If ornithologists coin the term 'Key sparrow' to name a new race of sparrows apparently discovered to be living in the Florida Keys, only later to find that all of the supposed exemplars observed were sophisticated animatronics planted by a glory-hungry birdwatcher, it seems we *would* say there are no Key sparrows (since the things observed were no birds at all), not that it turns out that Key sparrows are little robots" (Thomasson, 2007b, 49). If this were the case, the properties originally associated with the term would prevail over the discovery, casting doubt on Putnam's direct referentialism even as applied to natural kind terms.

Hence, Putnam's direct referential theory seems far from steady even when it comes to explaining the semantics of natural kind terms.

Consider now further disputes:

Dispute 5. 'Witch'/'Ether'. What do 'ether' or 'witch' apply to? Following Putnam's semantic theory, 'witch'—a social kind term—would apply to whatever shares the same "nature" as the paradigmatic witches. But what is the nature of witches? Critics of Putnam's thesis point out that the term seems to apply (if it applies at all) to whatever fits the definition connected to it. Since no one fulfills the description associated with the word 'witch', we have concluded that there are no witches (cf. Marconi, 2009, 233). A similar dispute concerns the name 'ether'—a natural kind term. Putnam would hold that 'ether' applies to the elastic, isotropic, homogeneous medium that fills the upper regions of space. However, since there is no substance—or rather no unified cause of the observed phenomena—it has been pointed out, once again, that this cannot be the case (cf. Marconi, 2009, 234). Indeed, the term was ultimately forsaken.

Other disputes. What does the term 'woman' apply to? Or 'parent'? Or 'race'? Do these terms have a specific reference that is externally determined, as Putnam suggests? Are they rather "socially constructed" concepts, whose reference cannot

---

[7] For a detailed discussion of the matter, i.e., that artifactual and social kind terms cannot have a directly referential semantic, see Olivero (2019).

[8] A proponent of direct referentialism may back up the foundation for all these cases, arguing that Putnam's theory applies to words for natural substances and species, provided that they were at least originally believed to designate a natural kind (see on this: Marconi, 2009, 132). However, this interpretation faces problems—as made evident by the case of 'ether' (see on this: Marconi, 2009, 234).

be defined only by some worldly essence (cf. Haslanger, 2017)? Does a term such as 'race' actually refer to something? Should we keep it, for it serves some important function, or should we forsake it because of its derogatory import (cf. Haslanger, 2012, 384)?

The preceding examples illustrate that Putnam's direct referential theory struggles also with accounting for terms whose supposed referents turn out to be nonexistent or conceptually incoherent, as well as cases in which reference, social function, and ethical considerations are so closely entangled. These persistent difficulties suggest that a different approach to semantics is needed. In addition, all the unsolved disputes sketched above, I will show (Sect. 10.2), are also (ongoing) metaphysical disputes. Traditionally, these are addressed within what Thomasson (2020a) calls a "heavyweight meta-metaphysical realism"—a view that does not seem very promising when it comes to attempting to solve them. Section 10.3 will briefly illustrate this approach and the problems it faces, as well as outline the alternative deflationary, pragmatic accounts that Thomasson proposes to overcome such challenges. I will then show how adopting this normativist framework leads us to read the disputes here at stake as "metalinguistic negotiations", namely, as debates about how the relevant words *ought* to be applied, rather than about how they are actually applied (Sect. 10.4). This alternative perspective, which I call a "Negotiating Approach", has several advantages—which I will illustrate in Sect. 10.5—most importantly its promise as a more viable and appealing attempt to (dis)solve these disputes.

10.2 (Some) Disputes in Metaphysics

As I have argued at length elsewhere (Olivero, 2018, 2019), when it comes to disputes like those described in the previous section, semantics and metaphysics are strictly interrelated. As I will show, these kinds of quarrels are not merely semantic; they are also metaphysical and modal. [continue from previous paragraph] If we quickly look back at the examples mentioned above, we can see that they do not simply concern what the terms at stake mean, how they are used, or what they refer to. Rather, they are disagreements about what the referents themselves in fact *are*—that is, about their nature and identity conditions. In this sense, such disputes are metaphysical. Moreover, because they are about what these referents *essentially* are—about what modal features they possess or may possess across possible worlds—they are also modal disputes. Let us now recall those examples, so that this becomes explicit.

Dispute 0. Twin-"water". Direct referentialism gained its popularity when applied to natural kind terms, for it seems plausible to think that 'water' applies to whatever shares a common *nature* with what we have paradigmatically indicated by that term (that is, H_2O) (cf. Putnam, 1973, 1975), and not to whatever matches a certain description of its (superficial) properties. This is because—as Putnam puts it: "Once we have discovered that water (in the actual world) is H_2O, *nothing counts* *as a possible world in which [the substance called] water isn't H_2O*" (1975, 233). What is at stake in the dispute is not merely what the term 'water' applies to; the dispute is about what water *is*, about what *nature* it has and may not have.

Dispute 1. Pencil-organisms. Putnam and Schwartz are not only arguing about how the term 'pencil' is correctly applied. That question seems to depend—as their arguments show—on what they believe the term's extension *is*, namely, on what *pencils actually are*. Putnam emphasizes that the fact that we would not call anything 'pencil' unless it shares the same nature as the paradigmatic pencil-organisms supports his claim that, as with natural kind terms, what governs reference is the deep structure or "nature" of the *actual* referents (whether or not we have epistemic access to that "nature"). This implies that the dispute over what the term 'pencil' refers to depends on the metaphysical and modal status of *pencils*. Schwartz (1978, 1980, 1983) argues, on the other hand, that this dispute about the nature of pencils is utterly irrelevant to understanding the semantics of these terms, because the application of artifactual and social kind words depends simply on their superficial features and characteristic functions. Pencils could be growing on trees or could be generated by lightning stricking the ground, and we would still call those objects 'pencils' so long as they have the same form and purpose. This also attests to the fact that—contrary to what it might initially be objected—Schwartz' semantic view on artifactual kind terms presupposes committments about the metaphysics of their referents.

Dispute 2. Pencil-shaped devices for control. Nelson and Schwartz are arguing about the "nature" of *pencils*. Nelson claims that if we found out that all our pencils were devices for control and manipulation distributed by aliens, we would indeed apply 'pencil' rigidly and indexically to those objects, for we would have discovered the *actual* nature of pencils. Schwartz counter-argues that this is not the case, since in the scenario imagined by Nelson, we would have simply discovered an additional property of pencils, rather than their essence (since, in his view, there isn't any) (cf. 1980, 1983). In this case as well—one may argue—we would probably start manufacturing "regular" pencils, that is, typical tools for writing and drawing. Either way, once again, the dispute ultimately revolves around what pencils (necessarily and essentially) *are*.

Dispute 3. Martian spies-pediatricians. Something similar occurs in the case of social kind terms. Schwartz, on the one side, and Putnam and Nelson, on the other, are disputing the "nature" of pediatricians. Against the latter authors, Schwartz insists that what seems crucial to defining a pediatrician is that she attended medical school, specialized in medicine for children, and so on. The fact that the agent in question also specialized in secretly gathering information about a country, person, or organization does not prevent her from also being a pediatrician. Moreover, the fact that she is not a human being does not prevent that either, as long as she can treat our kids (cf. Schwartz, 1978, 1980, 1983). The fictional case described by Putnam does not appear to paint a scenario in which we would have discovered a "nature" that replaces what we initially thought determined membership in the kind of pediatricians (cf. Schwartz, 1980, 193; 1983). Nonetheless, the dispute ultimately revolves around what pediatricians *are or may be*.

Dispute 4 and the like. Robotic-animals. Disputes about whether robotic-cats—i.e., cat-shaped robots remotely controlled from Mars—or Key Sparrows—i.e., bird-shaped robots initially believed by ornithologists to be a new species—are called 'cats' or 'Key Sparrows' are disputes about whether those objects *are* actually cats or birds, and about whether cats or birds *may* be robots.

Dispute 5 and the like. Fictional objects. Disputes about what 'witch', 'ether', and similar terms apply to are disputes concerned with the essence of the referents of these terms, i.e., about whether there is such an essence and what that may be.

Other disputes. Consider also those currently debated matters about how terms like 'gender', 'sex' (cf. Mikkola, 2012; Haslanger, 2017), 'fetus', 'race', or 'parent' (cf. Haslanger, 2009, 2012) refer. A debated question around these terms concerns whether the criterion that governs their application depends on what their referents *essentially* are, on what necessary features defines membership in their extension.

I have here shown how the semantic disputes described in Sect. 10.1 are ultimately disputes about what the referents of the considered terms are and can or cannot be, that is, about what they are necessarily and essentially. It seems clear, then, that to attempt to solve the disputes in semantics, we need to look more closely at the related disputes in metaphysics and modality. These disputes can be handled by two opposing approaches: a hyperrealist one or a normativist-deflationary one. In the following paragraph, I will compare these two approaches.

10.3 Heavyweight vis-à-vis Deflationary, Normativist Approaches

Traditionally, metaphysical disputes like those mentioned in the previous section are dealt with within what could be called, borrowing Thomasson's terminology, a "heavyweight meta-metaphysical realism". This approach "presents metaphysics as involved in discovering deep facts about the world that are not knowable by employing straightforward empirical and/or conceptual means" (Thomasson, 2020a, 438), but are rather epistemically inaccessible. Such a meta-metaphysical approach relies on what Thomasson defines as a "heavyweight modal realism" (2020b): a descriptive modal view which posits that there are modal facts in the world (e.g., water is H_2O) waiting to be discovered, and that these facts explain the truth of our modal statements (e.g., "Water is necessarily H_2O").[9]

That the disputes described in the previous section are framed within such a meta-metaphysical hyperrealist view is evident from the fact that all of them assume that there is an essence out there in the world (e.g., chemical composition, DNA, etc.) that defines what constitutes membership in the relevant kinds (e.g., the kind

[9]As Thomasson puts it: "[the metaphysician] care[s] about the real modal properties of things in the world [...]; the metaphysician sets out to discover these category of entity and to determine the modal features of entities of different categories" (2013, 144).

"water", "pencil", "Key Sparrow", and so on). Moreover, it is presupposed that such an essence may be inaccessible to us from an epistemic point of view, while still determining membership in those kinds. Putnam, for instance, specifies: "That pencils are artifacts is necessary in the sense of true in all possible worlds—metaphysically necessary. But it doesn't follow that it's epistemically necessary" (1975, 243). This is, in a nutshell, his hyperrealist view on the metaphysics of modality: there are deep facts in the world—that pencils are artifacts, for example—which make statements like "Pencils are (necessarily) artifacts" true. Our job—Putnam seems to imply—is to discover what these facts are when possible (cf. Thomasson, 2013, 2018, 2020b), even though we may never fully know (or discover) what they are.[10]

Furthermore, the use of a heavyweight metametaphysical realism to address the aforementioned disputes is reinforced by the observation that the metaphysical disputes outlined in the previous section underlie the semantic disputes presented in the first section. Those semantic disagreements revolve around Putnam's direct referential theory, which is a view that not only presupposes but requires a (hyper) realist theory of the metaphysics of modality to ground its account of reference—as shown by the already highlighted interconnection between semantics and metaphysics (cf. fn. 10).[11]

The other authors who have engaged in those disputes also embrace heavyweight metametaphysical and modal approaches. As the examples show (disputes 1 to 5), they engage in describing possible essences (or lack thereof)—i.e., identifying features metaphysically necessary albeit epistemically contingent—for the objects (or categories of objects) at stake.[12]

The heavyweight metametaphysical realist approach I have briefly described encounters several difficulties—as Thomasson points out (2013, 2016, 2020a, 2020b)—the most familiar of which are the epistemic mysteries about how the relevant guiding metaphysical facts could ever be discovered. Looking more closely at

[10]This heavyweight modal realism grounds Putnam's semantic theory. His direct referentialism is deeply and intimately connected to his hyperrealist metaphysical view. Indeed, Olivero argues elsewhere (2018, 2019) that there are two fundamental requisites for a term to be directly referential in Putnam's sense: (a) the term refers to things that have a *nature* (necessary features) possessed by all and only the members of the term's extension (that *nature* is what that determines the term's reference); (b) linguistically competent speakers can be ignorant of, or mistaken about such nature. This shows that Putnam's semantic theory cannot be considered separately from a realist theory of metaphysics and of the metaphysics of modality. I am aware that fully proving this point requires much more space. I have developed this argument further in another paper.

[11]As previously noted, Putnam's semantic view presupposes that the world is essentially full of deep or fundamental facts; we just may not have epistemological access to all of them—or perhaps to any of them. Still, these facts (e.g., that water is H_2O) are what determine our term's extensions (e.g., the extension of 'water'). That Putnam is committed to this view is further shown by the fact that the foundation of his "physical externalism" (i.e., his well-known direct referential theory) is scientific essentialism (cf. Liu, 2002). That is, the view that natural kind terms (in particular) refer to "natural objects [that] are marked by their natural boundaries" (2002, 392), boundaries that are to be discovered by scientists.

[12]A description of the debate about artifactual kind terms that shows further what stated can be find in Olivero and M. Carrara's "On the Semantics of Artifactual Kind Terms" (2021).

the examples discussed (disputes 1 to 5), both parties engaged in the disputes seem to believe that some modal facts in the world will ultimately decide which position is correct. It is just about discovering what those modal facts are. However, a question remains unanswered: how can we come to know them? This view leaves us appealing to facts that may, in principle, remain forever beyond our epistemic reach. And this is precisely what makes these disputes difficult to resolve within a purely descriptive framework. To spell out those difficulties in detail is not the interest of the present paper.[13] I am instead more narrowly concerned with confronting this traditional heavyweight account with valid alternative approaches to dealing with the disputes in question.

To bypass the difficulties into which the heavyweight accounts run, Thomasson proposes adopting alternative approaches both to modality and to meta-metaphysics (2013, 2016, 2018, 2020a, 2020b). Her alternative approach to modality, *Modal Normativism*, denies that modal statements are descriptive at all.[14] The basic function of talking about what is *metaphysically necessary* is not to try to describe modal features of the world. It is, rather, to provide a very useful way of expressing, conveying, or (re)negotiating constitutive semantic and conceptual rules in the object language.[15] Adopting Modal Normativism brings about several advantages (cf. Thomasson 2013, 2018, 2020b). One in particular is demystifying the epistemology of modality. On this alternative view, indeed, to acquire modal knowledge, we do not need to assume "that there must be a causal relation to the relevant facts or features to legitimate the claim to have knowledge of them" (Thomasson, 2020b, §7.3). Rather, we can simply consider the multiplicity of actual or imagined cases in which the terms would properly be applied or refused, to acquire a more explicit grasp of their underlying constitutive semantic rules.[16] This way, modal truths are knowable without resorting to anything mysterious (i.e., deep modal facts in the world). At the same time, we do not miss out on what seems so appealing in those

[13] For an accurate discussion of these criticisms, see Thomasson (2013, 2018, 2020a, 2020b *Intro* §1).

[14] Thomasson backs up her non-descripritivist approach to modality in the work of some early twentieth century analytic philosophers, such as M. Schlick (1918), L. Wittgenstein (1953), R. Carnap (1947), A. J. Ayer (1936), W. Sellars (1958), and G. Ryle (1949). She argues that their approach was mistakenly abandoned for the wrong reasons. Thomasson resuscitates it, making it a view that avoids the most important criticisms faced by those accounts (see Thomasson, 2020b).

[15] In Thomasson's own words: "On the normativist view, basic metaphysical modal claims do not have the function of tracking or describing special modal features of this world—or of describing other possible worlds. Instead, on this view, metaphysical modal language serves the function of expressing, teaching, conveying, or (re-)negotiating semantic rules (or their consequences) in particularly advantageous ways" (2018).

[16] In a way, this is also what Putnam is doing (cf. disputes 0, 1, 3, 4). However, the difference between Putnam's and Thomasson's underlying reasoning is substantial. From a modal normativist perspective, the intuitions that Putnam is calling on to support his thought experiments and actual examples would not serve to prove that there are modal facts in the world to appeal to in order to verify the truth of some modal statements (e.g., that water is necessarily H_2O). Rather, those intuitions play a major role in making explicit which semantic rules we follow when we use certain terms (e.g., natural kind terms).

rival descriptive accounts. For, on Modal Normativism, the relevant constitutive rules remain world-deferential, in the sense that, for example, the rule for 'water' prescribes that the term refers to "whatever has the same chemical structure as this (the paradigmatic sample)", leaving it to empirical discovery to find what that structure is (cf. Thomasson, 2020b, § 4.3).[17] Similarly, Thomasson's alternative deflationary approach to metametaphysics—that encompasses Modal Normativism—presents the same advantages, particularly that of demystifying the methods of metaphysics. For it is an approach that "relies on nothing more mysterious than (normative and descriptive) conceptual work—perhaps combined with straightforward empirical work—in answering those metaphysical questions that are well formed and fit to answer" (2020a, 437–438). This does not mean that such a "pragmatic (rather than metaphysical) methodology for normative conceptual work" (2020a, 440) is arbitrary, for it remains responsive to empirical and world-guided constraints (cf. Thomasson, 2020a, § 4.2).

A detailed discussion of all the advantages that these alternative accounts bring is also beyond the scope of this paper.[18] Here I am concerned with showing how embracing this deflationary, pragmatic framework leads us to see the disputes described in the previous sections (1 to 5) as "metalinguistic negotiations" (cf. Plunkett & Sundell, 2013; Plunkett, 2015; Thomasson, 2016). That is to say, they can be read as pragmatic disputes conducted in the object language about how the terms at stake *ought to be applied*, rather than as disputes about how they are in fact applied. The following section is dedicated to showing precisely that, while in the last section I illustrate the advantages that this pragmatic perspective entails.

10.4 A Pragmatic Approach to (Some) Disputes in Semantics

Adopting the described pragmatic methodology for prescriptive (and descriptive) conceptual work leads us to see several disputes in metaphysics, like the ones described in Sect. 10.2, as metalinguistic negotiations—namely, as disputes that, ultimately, are not about deep facts of the world, but are rather pragmatic disputes about how certain terms *should* be employed (cf. Thomasson, 2016, 2020a).

A metalinguistic negotiation is involved when speakers use a term with the pragmatic function of negotiating how (or perhaps whether) that very term *ought* to be used (cf. Plunkett & Sundell, 2013, 15). Imagine a 7-year-old girl in front of a dish with a side of boiled spinach and French fries. When asked to have her vegetables, the girl eats her fries and pushes the spinach aside. Her mom repeats the request, explaining she was referring to the spinach (since fries are not vegetables), but the girl is clear: she has already had vegetables. A metalinguistic negotiation has begun

---

[17] To that extent, it can be argued that Modal Normativism can fully account for the intuitions that favored the widespread appeal of direct referentialism.

[18] For a detailed discussion of the advantages of these normativist-deflationary accounts I defer to Thomasson (2013, 2018, 2020a, 2020b *Intro* §2).

alongside the nutritional one: no matter how hard her mother tries to explain the differences between healthy greens and fried food, the girl persists in advocating that fries *are* vegetables. She does not relinquish her position if presented with evidence, such as from books or dictionaries. Nor is she interested in the fact that other people agree with her parents on how the term is correctly applied.[19]

Similarly, the disputes reported in Sect. 10.2 may be read *not* as disagreements about possible discovered facts in the world, or about the ways the term in question is actually appropriately used, or even about the different ways the two discussants are using it. Rather, we may see the two parties involved as engaging in negotiations about how the relevant term *should* be applied—a reading supported by the observation that the disagreements often persist even after agreement on all empirical facts. By adopting the described alternative pragmatic framework, we can view all the disagreements at stake here not as disputes about what deep facts of the world might decide between the contenders involved, but rather as pragmatic disputes conducted in the object language about how the terms we use *ought to be applied.* This perspective seems to provide a more appealing account of why these disputes arise, as well as a more promising attempt to (dis)solve them—as the last section will further highlight. Let us now recall the relevant disputes so that this becomes evident.

Dispute 1. The pencil-organisms dispute. This dispute can be seen as a genuine metalinguistic negotiation, for it presents all the relevant hallmarks.[20] We can observe that Putnam and Schwartz agree on the relevant facts involved; nonetheless, Putnam asserts that, in the imaginary relevant scenario, pencil-organisms are pencils, whereas Schwartz claims that they are not. Even if we interpreted the disputants as using the term ‘pencil’ in different ways, they would still be debating something at the pragmatic level: whether ‘pencil’ *should* be applied to organisms or *should* be reserved for artifacts. In this light, Schwartz and those who disagree with the application of ‘pencil’ to the animated objects would be pragmatically advocating for a different use of that very term, because of the possible consequences for those organisms. Indeed, animal rights advocates would likely object to such a classification.

Dispute 2. The pencil-shaped devices for control dispute. We can imagine this dispute as one between the malevolent aliens who replaced our pencils with the devices of control and manipulation and common Earthlings. The malevolent aliens may argue that pencils-shaped devices for control *are* pencils not with the purpose of giving a competing description of a modal fact out there, but to *pragmatically*

---

[19] This and similar disputes are characterized by some hallmarks that, according to David Plunkett and Tim Sundell (2013), distinguish metalinguistic negotiations from other verbal disputes. In Thomasson’s interpretation, those hallmarks are: (a) the disputants may share the same knowledge or beliefs on the matter, and agree that no further discovery may resolve the debate; (b) even if the disputants agree on how the term is commonly used, the dispute does not end; (c) the dispute may persist even if the disputants acknowledge the fact that they are using the term differently (cf. Thomasson, 2016, 10).

[20] For relevant examples, see, for instance, Plunkett and Sundell (2013, 16, that is, the “Secretariat” example), or Thomasson (2016).

advocate for a certain application of that very term. The aliens would be negotiating a revision of the use of the term 'pencil' based on practical purposes and considerations. After all, their interest is to continue spying on and manipulate us through those devices. Their counterpart (the Earthlings), on the other hand, can maintain that 'pencils' *ought not* to be applied to the spying devices, precisely to preserve the Earthlings' rights to privacy, independence, and so on. In both scenarios, the negotiation will result in another practical consequence: as Schwartz points out (cf. 1983, 477), we would probably start manufacturing "real" pencils, making even more explicit the importance of the dispute.

Dispute 3. The pediatrician-Martian spy dispute. Those who maintain that 'pediatrician' applies also to Martian spies, as long as they graduated from medical school and are able to treat our kids, can be seen as actually advocating for a non-discriminatory use of the term (who are we to say that pediatricians have to be humans?). Their opponents, by contrast, could be pragmatically requesting a revision of the application of the term, based on the possible inauspicious import that the spying component would represent for our kids' lives.

Dispute 4 and the like. The robotic-animals disputes. These disputes, can be seen as negotiations similar to dispute 1, in which what is at stake are worldly issues, such as, for instance, what would happen if we were to call robotic-cats remotely controlled from Mars 'cats' and treat them as we treat regular animal cats. One may argue that, to a certain extent, calling the robotic-cats 'cats' does not make much difference in our lives, since they would perform the same "function" as the animals do. Still, what is at stake is a negotiation about how the term ought to be applied—a negotiation that depends on a *decision* based on pragmatic purposes, interests, or ethical concerns (cf. Haslanger, 2016), rather than *on uncovering* deep, fundamental facts about the world (cf. Thomasson, 2020b, § 4.1).

Dispute 5 and the like. The fictional objects disputes. We may see negotiations taking place not only over how some words should be applied but also over whether they *should be applied at all.*[21] Think about the term 'witch'. One may insist on its disuse, based on the negative conveyed meaning and social consequences it once carried. Recall the time when the term was actually used to discriminate against—and even burn—the women labeled with it. Utterances such as: 'Witches do not exist" may be used to alter the norm for using the term, in light of its pragmatic impact on our way of life. In this sense, utterances of trivial truths or falsehoods like these (e.g., "Witches do not exists", "Robotic cats are not animals", "Pencils on Twin Earth are not pencils", and so on) may have "a deeper pragmatic point: to press for ways the very linguistic/conceptual scheme used in expressing the obvious truths or falsehoods is to be used" (Thomasson, 2016, 12). "For regardless of whether or not we see the speakers as literally asserting conflicting propositions, their real dispute arises at the level of what it is pragmatically communicated" (2017, 111).

[21] For other relevant examples, see Thomasson (2016, 13–14; 2017, 113; 2020a, § 2).

Other disputes. Disagreements about the application of terms like 'gender', 'fetus', 'parent', and so on seem clearly to be disputes that cannot be sorted out by merely looking at the world. They are plausibly interpretable as metalinguistic negotiations about how these terms should be applied, about what they ought to refer to. For, again, determining these terms' reference seems to be a matter of decisions based on interests, purposes, ethical concerns, and so on. These disputes highlight that how we choose to apply words can have far-reaching effects, revealing commitments and attitudes that transcend language and have little to do with deep modal facts.

If this interpretation is convincing, it suggests adopting a "Negotiating Approach", which allows us to read the disputes at stake as metalinguistic negotiations. Within this framework, the primary focus is the negotiation of meaning and reference—a negotiation based on both pragmatic and empirical factors. Such a perspecive has several advantages over competing semantic views (e.g., Putnam's direct referentialism), as I will discuss in the following section.

10.5 Advantages of a Negotiating Approach

As argued at length (Sects. 10.3, 10.4), if we take a deflationary, normativist, pragmatic approach to dealing with the disputes at stake in the present paper, we no longer view these disputes as instances in which there are modal features of the world that outstrip our conceptual resources and ability to know them. Therefore, one significant advantage of adopting such an alternative framework is that it demystifyies the epistemology of the metaphysics of modality—a considerable advantage over its descriptive competing accounts.

This, however, does not mean that such a framework is arbitrary or conventional in any sense, for—as highlighted in Sect. 10.3—it has a way to accommodate empirical discoveries: such a framework is certainly world-responsive. For instance, in a scenario in which all of our cats turned out to be robots, what is in question would not be this very fact, but rather what we *should* do with the term 'cat': should we keep it to refer to those objects or should we abandon it? Within a Negotiating Approach that relies on such a pragmatic framework, we gain the advantage of a more plausible explanation of what we do. The answer to that question *does not* (and *should not*) depend on anything mysterious out there, but rather on our interests, concerns, purposes, and so on, without, nonetheless, neglecting the empirical factors.

This concerns not only allegedly artifactual kind terms (or, more generally, terms related to the "cultural realm", so to speak). Consider the following scientific disputes. The term 'red panda' has been retained despite the fact that its referents do not exhibit the properties originally connected to it. Red pandas belong, in fact, to a peculiar family, having more features in common with felines than bears. By contrast, the term '*Equus johnstoni*' applied to a species of African ungulates was abandoned. After it was found that those animals were not related to horses but giraffes, the original term was replaced with '*Okapia johnstoni*' (cf. Marconi, 2009, 130). A Negotiating Approach has a plausible explanation for these and similar cases (cf. disputes 1, 4, 5 Sect. 10.4), whereas Putnam's direct referentialism struggles. What is the modal fact in the world that made us keep the term 'red panda' but not '*Equus johnstoni*'? It seems that on these occasions, namely in the face of empirical discoveries in which things turned out not as expected, we have to *negotiate* what to do with our terms and original beliefs about their reference: a negotiation based on factors beyond just empirical evidence.[22]

With its world-responsiveness, such a Negotiating Approach maintains the advantage of also accounting for the intuitions that are "externally driven" and that motivated the wide propagation of direct referentialism and the strong realist view behind it. A Negotiating Approach still provides a persuasive explanation of why we would not say that a liquid with a different microstructure *is* water, or that iron pyrites *is* gold (cf. Putnam, 1975), or why we would, instead, agree that black swans *are* swans[23] or that 'Covid-19' is a directly referential expression. On the other hand, within this alternative view, we are also able to adequately account for the disputes or conflicting intuitions that (may) arise about these words and their application. For example, a Negotiating Approach can readily account for the fact that we call 'water' the liquid we drink, even if it is not primarily composed of H_2O molecules (cf. Weisberg, 2006).

Moreover, the alternative framework outlined here carries a further significant advantage of seeing these disputes not as merely verbal or nonsensical—as they would be seen within traditional deflationist approaches (cf. Thomasson, 2016, 2017). By qualifying the relevant disputes as *prescriptive* discussions about how the terms involved *should* be used, we gain the great payoff of being able to explain why they may be worth having. "For how we use words *matters,* given their relations to other aspects of our conceptual scheme, and to our non-verbal behavior." (Thomasson, 2016, 9). As the examples show, other issues (e.g., ethical) are at stake along with the semantic ones. A Negotiating Approach can therefore also account for the importance of these disputes and why they arise in the first place.

---

[22] Carnap already noted that the change in meaning of 'fish' brought about by zoologists "was not a correction in the field of factual knowledge but a change in the rules of language; [although] this change, it is true, was motivated by factual discoveries" (1950/1962, 6) [cit. in Thomasson, 2020b, § 4.1].

[23] This example refers back to one described by Tyler Burge in his "Individualism and the Mental" (1999), which served, among other examples, to explain his theory now known as *social externalism.* The example specifies that, if the linguistic norm adopted by a givem community claimed that 'swan' means (or is synonymous with) 'white swan', the term 'swan' could not be extended to the black birds, regardless of the fact that both types of birds belong to the same species. According to social externalism, indeed, what determines the meaning and the reference of our words is solely the semantic norm adopted by the linguistic community in which these words are employed—no empirical fact or discovery is to be taken into account. In this respect, social externalism is effectively a form of masked descriptivism (cf. Olivero and Carrara, 2021). For a detailed discussion of the relevant differences between Putnam's physical externalism and Burge's social externalism, see Liu (2002).

Furthermore, within the Negotiating Approach I proposed here, meaning maintains its characteristic feature of being a variable. Putnam already held this thesis. However, in Putnam's account, the value of that variable is determined solely by the nature of the physical environment (cf. Liu, 2002), whereas on the alternative framework offered, that value is negotiated according to pragmatic and empirical factors (nature, authority, ethics, etc.). Seeing this component of meaning is also crucial for its further pragmatic, social significance: meaning can be "engineered", namely, it can be revised or replaced based on pragmatic concerns (cf. Plunkett, 2013, 2015; Cappelen et al., 2020; Thomasson, 2020a). For instance, on this alternative view, we could see that a word like 'parent' cannot serve a good purpose (for several pragmatic and ethical reasons) if it refers—as Putnam's direct referentialism would hold—only to the immediate biological progenitor. Rather, 'parent' would serve a better function if its reference is not limited to that, but also covers other socially constructed features associated with that concept (cf. Haslanger, 2009, 2012). Revising and adjusting meaning in this sense has considerable pragmatic and social relevance.

10.6 Conclusion

I have here drawn attention to some recent disputes in semantics and metaphysics, disputes that arise particularly around Putnam's directly referential semantic theory and the hyperrealist modal view on which it relies. Putnam's accounts have not been successful in solving these disputes, as several objections have been raised. I have illustrated how such disagreements are better addressed by adopting a deflationary, normativist, pragmatic framework. Within such a framework, we can see these disputes through the lens of what I labeled a "Negotiating Approach", namely, one that sees them as metalinguistic negotiations about how the terms at stake ought to be applied.

I have shown that such a Negotiating Approach has several advantages over its competitors traditionally employed to untangle those disputes. On the one hand, it avoids the mysterious epistemology that comes with heavyweight traditional accounts. On the other hand, via the shift of attention to a normative, pragmatic level, it provides a more viable and appealing theory of why those disputes arise in the first place, as well as a more attractive attempt to (dis)solve them. For other issues are at stake (ethical, legal, political, etc.) along with the semantic ones. The disputes in question shall not be dismissed as merely verbal or nonsensical and will be recognized as very much worth having.

Shifting the attention, though, does not mean that we should disregard possible relevant empirical facts or discoveries, as it happens with traditional deflationary approaches. For the pragmatic approach presented here takes these facts into account and accommodates them. We then gain an improved account of the significance of the disputes in question, while still preserving the persuasive points that motivated competing accounts (e.g., direct referentialism). Lastly, by adopting this Negotiating Approach, we can see that meaning is a variable, whose value ought to be negotiated according to pragmatic and empirical factors. This paves the way for engineering meaning, namely, for normatively revising and improving it when needed, in accordance with the relevant factors—a powerful tool for potentially improving society.

References

Abbott, B. (1989). Nondescriptionality and natural kind terms. *Linguistics and Philosophy, 12*, 269–291.

Ayer, A. J. (1936/1952). *Language, truth and logic*. Dover.

Burge, T. (1999). Individualism and the mental. *Midwest Studies In Philosophy, 4*, 73–121.

Capellen, H., Plunkett, D., & Burgess, A. (Eds.). (2020). *Conceptual engineering and conceptual ethics*. Oxford University Press.

Carnap, R. (1947). *Meaning and necessity: A study in semantics and modal logic*. The University of Chicago Press.

Carnap, R. (1950/1962). On explication. In *Logical foundations of probability*. The University of Chicago Press.

Olivero, I. & Carrara, M. (2021). On the Semantics of Artifactual Kind Terms. *Philosophy Compass, 16*(11), 1–13.

Elder, C. (1989). Realism, naturalism, and culturally generated kinds. *The Philosophical Quarterly, 39*, 425–444.

Elder, C. (2007). On the place of artifacts in ontology. In E. Margolis & S. Laurence (Eds.), *Creations of the mind: Essays on artifacts and their representation* (pp. 33–51). Oxford University Press.

Frege, G. (1892). Über Sinn und Bedeutung. *Zeitschrift für Philosophie und philosophische Kritik, 100*, 25–50.

Haslanger, S. (2009). Family, ancestry and self: What is the moral significance of biological ties? *Adoption & Culture, 2*, 91–122.

Haslanger, S. (2012). *Resisting reality: Social construction and social critique*. Oxford University Press.

Haslanger, S. (2016). Theorizing with a purpose: The many kinds of sex. In C. Kendig (Ed.), *Natural kinds and classification in scientific practice* (pp. 129–144). Routledge.

Haslanger, S. (2017). The sex/gender distinction and the social construction of reality. In A. Garry, S. J. Khader, & A. Stone (Eds.), *The Routledge companion to feminist philosophy* (pp. 157–167). Routledge.

Kaplan, D. (1989). Demonstratives. In J. Almog, J. Perry, & H. Wettstein (Eds.), *Themes from Kaplan*. Oxford University Press.

Kornblith, H. (1980). Referring to artifacts. *Philosophical Review, 89*, 109–114.

Kornblith, H. (2007). How to refer to artifacts. In E. Margolis & S. Laurence (Eds.), *Creations of the mind: Theories of artifacts and their representation* (pp. 138–149). Oxford University Press.

Liu, J. (2002). Physical externalism and social externalism: Are they really compatible? *Journal of Philosophical Research, 27*, 381–404.

Marconi, D. (2009). Being and being called. *The Journal of Philosophy, 106*, 113–136.

Marconi, D. (2013). Pencils have a point: Against general externalism about artifactual words. *Review of Philosophy and Psychology, 4*, 497–513.

Mikkola, M. (2012). How essential is gender essentialism? *Symposia on Gender, Race, and Philosophy, 8*.
Nelson, J. (1982). Schwartz on reference. *The Southern Journal of Philosophy, 20*, 359–365.
Olivero, I. (2018). Putnam on artifactual kind terms. *Review of Philosophy and Psychology, 9*, 197–212.
Olivero, I. (2019). Function is not enough: An externalist defeat for artifactual and social kind terms. *Grazer Philosophishe Studien, 96*, 105–129.
Plunkett, D. (2015). Which concepts should we use?: Metalinguistic negotiations and the methodology of philosophy. *Inquiry, 58*, 828–874.
Plunkett, D., & Sundell, T. (2013). Disagreement and the semantics of normative and evaluative terms. *Philosophers Imprint, 13*(23), 1–37.
Putman, D. (1982). Natural kinds and human artifacts. *Mind, 91*, 418–419.
Putnam, H. (1973). Meaning and reference. *The Journal of Philosophy, 70*, 699–711.
Putnam, H. (1975). The meaning of 'meaning'. In *Mind, language and reality, philosophical papers* (Vol. 2, pp. 215–271). Cambridge University Press.
Russell, B. (1905). On denoting. *Mind, 56*, 479–493.
Ryle, G. (1949). *The concept of mind*. Hutchinson.
Schlick, M. (1918). *Allgemeine Erkenntnislehre*. Springer.
Schwartz, S. (1978). Putnam on artifacts. *Philosophical Review, 87*, 566–574.
Schwartz, S. (1980). Natural kinds and nominal kinds. *Mind, 89*, 182–195.
Schwartz, S. (1983). Reply to Kornblith and Nelson. *The Southern Journal of Philosophy, 21*, 475–479.
Sellars, W. (1958). Counterfactuals, dispositions and the causal modalities. In H. Feigl, M. Scriven, & G. Maxwell (Eds.), *Concepts, theories and the mind-body problem, minnesota studies in philosophy of science* (Vol. 2, pp. 225–308). University of Minnesota Press.
Thomasson, A. L. (2003). Realism and human kinds. *Philosophy and Phenomenological Research, 67*, 580–609.
Thomasson, A. L. (2007a). Artifacts and human concepts. In E. Margolis & S. Laurence (Eds.), *Creations of the mind: Theories of artifacts and their representation* (pp. 52–73). Oxford University Press.
Thomasson, A. L. (2007b). *Ordinary objects*. Oxford University Press.
Thomasson, A. L. (2013). Norms and necessity. *The Southern Journal of Philosophy, 51*, 143–160.
Thomasson, A. L. (2016). Metaphysical Disputes and Metalinguistic Negotiation. *Analytic Philosophy, 58*, 1–28.
Thomasson, A. L. (2017). In G. D'Oro & S. Overgaard (Eds.), *What can we do, when we do metaphysics?* (pp. 77–100). Cambridge University Press.
Thomasson, A. L. (2018). How can we come to Know Metaphysical Modal Truths? (in Synthese special issue: *New Directions in the Epistemology of Modality*).
Thomasson, A. L. (2020a). A pragmatic method for conceptual ethics. In H. Capellen, D. Plunkett, & A. Burgess (Eds.), *Conceptual engineering and conceptual ethics* (pp. 435–458). Oxford University Press.
Thomasson, A. L. (2020b). *Norms and necessity*. Oxford University Press.
Weisberg, M. (2006). Water is not H_2O. In D. Baird, E. Scerri, & L. McIntyre (Eds.), *Philosophy of chemistry. Boston Studies in the Philosophy of Science* (Vol. 242, pp. 337–345). Springer.
Wittgenstein, L. (1953). *Philosophical investigations*. Blackwell.

Chapter 11
Replies to Authors

Amie L. Thomasson

Abstract In this text, I reply to the other papers this volume, which include discussions of my work on fiction, metaontology, and modality. Regarding my artifactual theory of fiction, I respond to issues about the reference of fictional names, and discuss whether the artifactual theory of fictional characters I develop can be extended to characters in films, or to the theoretic entities of scientific theories. On metaontology, I defend my metaontological views about the ontology of art, and defend an easy ontology of facts, while also discussing what ontological discourse does for us. Finally, on the topic of metaphysical modality, I respond to accusations that my views about metaphysical modality unduly rule out substantive metaphysical debates, and discuss whether these views can be extended to handle counterfactuals and talk of possible worlds, and how they may be relevant to debates about the meanings of artifactual and natural kind terms.

This volume began at a conference on my work held in Granada, Spain in December 2019. I am extremely grateful to all the philosophers who participated there, and/or contributed here, for taking the time to engage so deeply with my work and for providing so much to think about. I am also grateful to Maria Garcia-Encinas and Fernando Martínez Manrique for organizing the conference, to Otavio Bueno for suggesting the idea of a volume based on the work presented there, and to Mirco Sambrotta for all his work in editing this volume.

The papers collected here address work spanning my career to date—from my early work on fiction 25 years ago, to my recent work on modality and metametaphysics. I will address and reply to their contributions below, roughly in time-order of the work discussed. While some of the papers here raise challenges for my views,

A. L. Thomasson (✉)
Department of Philosophy, Dartmouth College, Hanover, NH, USA
e-mail: Amie.L.Thomasson@dartmouth.edu

M. Sambrotta (ed.), *Metaphysics Today: In Conversation with Amie Thomasson*, Synthese Library 512, https://doi.org/10.1007/978-3-031-98468-6_11

others apply or extend the views in useful and important ways. In either case, engaging with these critical discussions of my work has been both thought-provoking and valuable to me. I hope readers, too, will find that reading these articles and replies enhances their understanding of the positions I defend, their challenges, their resources for defense, and their broader implications for other areas of philosophy—and sometimes, for other areas of life.

11.1 Acero on Fiction

Juan Acero takes us back to some of my earliest work—*Fiction and Metaphysics* (Cambridge University Press, 1999)—in which I develop an 'artifactual theory' of fiction. When I first wrote *Fiction and Metaphysics,* there were two major approaches to fiction. One (anti-realist) approach holds that the names of fictional characters never refer; instead, all apparent references to fictional characters are to be paraphrased away (as on traditional Russellian, Fregean, or Rylean approaches) or taken as within the context of a pretense or game of 'make-believe' (Walton, 1990). The second (neo-Meinongian) approach holds that the names of fictional characters *do* refer—to non-existent or Platonistic abstract objects, which are not created, and are individuated by the properties they are attributed in the stories (Parsons, 1980; Zalta, 1983). In contrast to both of these approaches, I argued that there *are* fictional characters, but that these are a kind of abstract cultural artifact, created in a certain time and context, and capable of change and destruction. In these ways, I argued, they are not unlike works of literature, works of music, laws of state, and other abstract cultural artifacts.

Acero argues that this approach runs into what he calls the 'Equivocality Problem'. The problem is this: Names for fictional characters get used in many different contexts.[1] In some (which I call 'external' contexts), we speak from the perspective of a critic or literary historian, where we might speak of Harry Lime, say, as a character created by Graham Greene in *The Third Man*. In other contexts (which I call 'internal contexts') readers discuss the *content* of a work of fiction, e.g. saying "Harry Lime traffics in diluted penicillin". Acero claims that to handle both kinds of discourse, I must treat "Harry Lime" as two names in disguise—on my view, he alleges, "'Harry Lime' is a two-symbol, equivocal expression" (Acero 32).

And this, he argues, leads to other problems. For what look like sound inferences turn out not to be sound, given the equivocation. For example, consider the following inference:

1. Harry Lime is the character created by Graham Greene in The Third Man
2. Harry Lime traffics in diluted penicillin

[1] In later work (2009, 11) I distinguish four different forms of fictional discourse: fictionalizing discourse (discourse within works of fiction), nonexistence claims, internal discourse (by readers about the content of works of fiction), and external discourse by readers and critics (about characters as characters, their sources, origins, place in literary history…).

3. The character created by Graham Greene in The Third Man traffics in diluted penicillin

For in (1) (Acero says, correctly) I treat 'Harry Lime' as referring to an abstract artifact; but in (2) (he suggests that my view says) 'Harry Lime' fails to refer (Acero 13).

Anyone who works on fictional discourse must do *some* paraphrasing in order to avoid outright contradictions, say, between saying that Frankenstein's monster was a creation of Mary Shelley, and that Frankenstein's monster was created by Dr. Frankenstein (see my 2003a). No one can just take all statements about fictional characters at face value.

But Acero oversimplifies my way of handling internal claims such as (2). As he notes, I do treat such internal discourse about the content of works of fiction as implicitly in the context of an 'according to the story' or pretense operator (this enables us to avoid contradictions like the one just mentioned).[2] But questions remain about how to read the scope of the 'according to the story' operator. That is, we could read the name 'Harry Lime' as *outside* the scope of the story operator:

2.1 Harry Lime is such that (according to the story, he traffics in diluted penicillin).

Or we could read name as inside the scope of the story operator:

2.2 According to the story (There is some x such that x is called 'Harry Lime' and x traffics in diluted penicillin)

If we take the first option, we can read the name 'Harry Lime' in 2.1 as referring to an abstract artifact—just as in (1), and yet we avoid contradictions between what is *said in the fiction* and what is *really the case,* since what is said in the fiction is only true *according to the story.* We also keep a parallelism in reading fictional discourse about real people, versus about fictional characters. If we take the second option, we need not take the name 'Harry Lime' in that context to be referring.

Acero claims to be discussing only the version of my view from *Fiction and Metaphysics* (1999), but there I embrace the first option, and take 'Harry Lime' to refer to the abstract artifact, and to report on what is true about it, according to the story.[3] This avoids the 'equivocality problem', since the name in each case refers to an abstract artifact.

[2] A pretense operator would also do the job—I am not particularly committed on which way to go.

[3] I was somewhat less explicit about the scope issue back in my (1999) than I would be now. But I do speak of predications in fictional contexts as *parallel* regardless of whether they are about real individuals or fictional characters (1999, 107). Internal statements "are not literally true, although they describe states of affairs that, according to the relevant stories, do obtain regarding the fictional character Hamlet and the real people Nietzsche and Freud" (1999, 107). I also accept there that authors performatively create fictional characters through their writing (1999, 12–14) and then refer back to them, ascribing them further properties. I also say there that authors of sequels refer back to extant characters and ascribe them new properties (1999, 67). I was more explicit about the choice point in reading the scope my (2003a, Sect. 2.1). Please see that text for fuller discussion.

But taking this approach does leave me with a different problem, raised by Stephen Yablo (1999): that it entails that authors are ascribing category-inappropriate properties (like smoking a pipe, or in this case, trafficking in diluted penicillin) to abstract objects.

For these reasons, I later (by the time of writing my 2003a) came to adopt the second option.[4] That is, I came to hold (roughly following Schiffer, 1996) that in *writing* fiction (engaging in *fictionalizing* discourse), authors aren't referring back to fictional characters, but rather *enabling* reference to them, by engaging in a kind of extended *de dicto* pretense that 'there once was a man, such that he…', which functions roughly as a performative that enables future readers to refer back to the *fictional character* as an abstract artifact created therein. (Those who engage in internal discourse about the contents of works of fiction may then be seen as reporting on the contents of this *de dicto* pretense.) This solution has the benefit of not treating authors as ascribing categorically inappropriate properties to abstract artifacts. As I acknowledge (2003a), it also has some costs of its own—for example, that it threatens to undermine the parallelism between fictionalizing discourse about real people versus fictional characters,[5] and complicates the story of how an author may write a sequel 'about' the same character.

Nonetheless, that doesn't mean that we treat the name 'Harry Lime' as *equivocal,* as merely 'two names disguised as one', as Acero alleges. For (*pace* Acero) the two uses of the name (in the author's fictionalizing discourse, and in external discourse) are *not* unrelated, mere accidental uses of the same letters, as in 'bat' (animal) and 'bat' (baseball), or like '4' being used to refer to both a number and a mountain. Instead, the fictionalizing uses *create* the character (abstract artifact) and *entitle* us to later refer back to the abstract artifact (in external contexts). To show the connection, we use the same name that was used in the original *de dicto* pretense (in internal discourse, we implicitly report back on the content of the original pretense). So, despite what Acero says, I *do* allow that fictional names "used in the context of fiction hold an intimate relation to their uses in the real context" (Acero 15).

It is also (*pace* Acero 16) not so strange to allow that the same name may be used in different (but related) ways to (attempt to) refer to things of different metaphysical categories. As I argue later (2010), in order to handle nonexistence claims involving fictional names, we need in any case to allow that fictional names may be associated with two different *categories* of (presupposed, assumed, or intended) referent. The nonexistence claims most often considered are those that we might make to correct someone who has mistakenly taken a fictional name for a name of

---

[4] For the record, though, I reject Acero's claim that I have 'abandoned a substantial part' of my former theory (though I have accepted this alteration in reading the scope of one kind of fictional discourse). I also want to note that the deflationary aspect of my work post-*Fiction and Metaphysics* is not a radical turn onto a different path, but merely makes explicit and more fully develops some themes that were already implicit in the phenomenological tradition, and in my own earlier work inspired by it. For connections between these ideas, see "How it all hangs together" (2023a).

[5] Depending, of course, on how one analyzes fictional discourse about real individuals.

a real person. To the person who wants to find Sherlock Holmes to help solve a crime, we might say 'Sherlock Holmes doesn't exist', or 'There's no such person—it's just a fictional character'. But we can also make claims of existence and nonexistence about *fictional characters,* for example if (in response to claims there are no black characters in the Harry Potter books) we say, "On the contrary, such characters do exist--there's Angelina Johnson, for example". To handle such cases (I argue (2010)) we need to allow that the same name may be associated with different sets of application conditions—say, those for fictional characters (which can be fulfilled by the name being used appropriately in a work of fiction), versus those for people (which would have to be fulfilled by a human being). But this is not so strange as Acero suggests. For there are many cases in which we can use what is in some sense the same term (same letters and pronunciation, related uses) to attempt to refer to things of different ontological categories—and this happens especially with terms for abstract artifacts. For example, "*The Third Man* is on the second shelf down" and "*The Third Man* is Graham Greene's most iconic tale"; or "The Affordable Care Act is nearly 1000 pages long" and "The Affordable Care Act expands the Medicaid program", or "Beethoven's Ninth Symphony was interrupted by a fire alarm" and "Beethoven's Ninth Symphony has four movements".[6]

The short response to the central problem Acero raises, then, is that it's *not* obvious that I treat the name 'Harry Lime' differently in (1) and (2); this depends how we read the scope of the story operator. In *Fiction and Metaphysics* (which Acero claims to limit his discussion to) I read the name as referring to an abstract artifact in both cases. I readily grant that there are other options, and that there are trade-offs to be made whichever way one goes—I ask only that they be explicitly acknowledged and clearly weighed up. In considering the options, I later (2003a) came to adopt the alternative view—that we should take (2) as giving a *de dicto* report on what a story says, where the story doesn't refer to an abstract artifact but *enables* reference to one (because engaging in such forms of literary pretense are *all it takes* to count as creating a fictional character) (see my 2003b). But this, too, doesn't leave such names as merely 'equivocal', or as 'two names in disguise', and does draw crucial connections between these two uses. It also seems to me to aptly report on some of the complexities in the relations among various different uses of fictional names—differences that are part of what makes handling fictional discourse so complicated and challenging.

11.2 Otavio Bueno on Fiction and Film

Otavio Bueno raises fascinating issues about whether the artifactual theory of fiction, developed with literary characters in mind, would carry over smoothly to talk about the fictional characters that appear in films.

[6] The same category shifting ability would go also for names of colors (and color samples), for names of mathematical proofs, (types of) car or other invention, and many more cases.

He makes two major points: first, there seem to be distinctive barriers to treating the characters portrayed in films as abstract artifacts. For it seems that, on my view, one should say that we *see* fictional characters by watching a film (or by seeing images on the screen), but "given fictional characters' abstract nature, how can one possibly *see* them?" (6). For "only objects that are in space and that reflect light can be seen" (6).

Second, he argues that, in any case, there is no need to accept that there are fictional characters (conceived of as abstract artifacts or otherwise). For a thoroughgoing empiricism can enable us to:

> remain largely agnostic about several issues that more substantive metaphysical views aim to settle—including the existence of fictional characters and abstract objects more generally. Settling these issues is ultimately not required to make sense of film or literature as our literary and cinematic practices are largely undisturbed by the way in which these metaphysical issues are resolved (9)

For we can say that we understand a film, including what is supposed to be real or imagined in the content of the film, by means of seeing images and engaging in perceptual imagination—taking the images to 'stand for' fictional characters, and imagining that there are the relevant characters and events. On this view, we can account for how we follow a narrative in a film without the need to take any stand on whether the characters apprehended exist—their "existence is not required to make sense of motion pictures" (10).

These are great questions to raise. And neither in *Fiction and Metaphysics,* nor elsewhere in my work, did I address questions about how or whether the artifactual theory of fictional characters might (or might not) carry over to characters that appear in films.

Let me first address the question of how (or whether) one could *see* the characters in films. Bueno assumes that, if such characters are indeed *abstract* (albeit, abstract *artifacts,* as I have it), we could not see them. And he seems to assume that if one accepted that there are cinematic fictional characters, these must be what we see when we view a film. Let me consider these issues in turn, which lead to two different possible lines of response.

The thought that we cannot (in any sense) see abstracta arises from thinking of abstracta on the classic Platonist model. Numbers and Platonic forms are the standard paradigms of abstracta, and they are supposed to be outside of space and time, and not things we can perceive with our senses. But once we shift to thinking in terms of abstract artifacts, it is not so clear that this carries over. For part of the point of the abstract *artifact* model is to note that many everyday entities we care about—including not only fictional characters, but also literary works, laws of state, symphonies, etc.—do not fit the Platonistic paradigm, even though they also are not concrete objects. Regrettably, I do not have a theory of perception or a detailed phenomenology of film to offer; that would require a great deal of additional work. But I do argue in *Fiction and Metaphysics* (Chap. 4) that we can extend the causal theory of reference to allow that we can make 'quasi-indexical' reference to fictional characters and other dependent abstracta "via the spatiotemporal concreta in which they are exemplified or represented or on which they depend" (1999, 52). So similarly, it would seem to make sense to allow that we can in some sense perceive (or better: quasi-perceive) some abstracta. We can, for example, see the color Hooker's Green Light through perceiving instances of it, and can hear symphonies via hearing performances of them. Along these lines, there is room for allowing a sense in which we see (or perhaps better: quasi-see) the fictional characters represented in film. Of course, this quasi-seeing must be somewhat different from ordinary visual perception—in ways phenomenologists have explored. As Roman Ingarden writes, "With justification one could say… that the things and people presented in the film are not simply imagined, but rather really seen by the spectator, although it is not a matter of a straightforward perceptual seeing" (1989, 326). Seeing the things presented in a film does, like ordinary perception, require our eyes being affected by light. But unlike ordinary visual perception, we are not free to move about to see what Sponge Bob looks like from behind, or what shape Mia Wallace's toes are—there are, as Bueno notes, places of indeterminacy in the perceived objects. In watching a film, we also cannot confirm what we take in visually by reaching out with touch, smell, or taste. The experience of time is different as well, as Sartre details in Part IV of his *Psychology of Imagination.* It would take far more detailed work in phenomenology to work out the differences fully (and much of this phenomenologists such as Edmund Husserl, Roman Ingarden, and Jean-Paul Sartre have worked on). But the short point here is to note that there is not a principled barrier to saying that in some sense fictional characters can (unlike numbers) be quasi-perceived in a way that does require the use of the eyes, as well as the imagination. (And in ways that would draw the contrast between how we see *some* characters in a movie, but not others (as, in "Charlie's Angels", we famously see the three women detectives, but never see Charlie). In fact, the interesting discussions Bueno introduces of perceptual imagination may be helpful in articulating what is involved in this sort of 'quasi-perception'.

But even if we allow that there is a kind of quasi-perception involved in viewing a film (and can see that there are fewer barriers to allowing something like perception of abstract *artifacts*), it remains an open question whether we should say that, in viewing the film, we are quasi-perceiving the relevant *abstract artifacts.* There is a parallel to the issue I discuss in the reply to Acero above about literary fiction: in writing a work of fiction, should we say that the author is saying things *about an abstract artifact,* or merely that they are pretending to assert things about a person, *in virtue of which* we are *entitled* to say that an abstract artifact (the fictional character) is created? As I mentioned there, while in *Fiction and Metaphysics* I took the former approach, in my (2003a) paper (2003a), (following Schiffer, 1996) I lean in favor of shifting to the latter view. I am inclined to run a parallel approach for films: saying that, in watching a film, we simply engage in perceptual imagination (without saying that the abstract artifacts are (quasi-)seen), while insisting that nonetheless *there* are fictional characters in films, and that we can refer to them in engaging in film history, criticism, and the like. As will become clear below, that leaves intact the primary reason for saying there are fictional characters: to make sense of (external) critical and historical discourse about them.

That brings us to Bueno's second claim: that a thoroughgoing empiricism can enable us to make sense of films, and our experience and interpretation of them, while remaining agnostic about what he calls "more substantive metaphysical views" about the existence of fictional characters and abstract objects more generally (9).

I actually agree that one can make sense of our experience and interpretation of films—or works of literature for that matter—without having to say that there *are* fictional characters (whether literary or cinematic). In *Fiction and Metaphysics* I distinguish two types of discourse: discourse in 'fictional contexts'—about what goes on in the work (here we speak of Anna Karenina and the like as (if they were) *people,* doing certain things, interacting with others, in a fictional world...); and discourse in 'real contexts'—where we speak from the external perspective of the literary critic or historian about the characters *as characters,* created in a certain time and context, with certain sources or influences, picked up in later stories, etc.). It was always the discourse *in real* (external) contexts that, for me, provided the core arguments for saying 'there are' fictional characters. For all the internal talk and experience, say, of Anna Karenina having an affair, lying on the train tracks, etc., can be handled perfectly well by treating it as talking about what is true *according to the story,* or what we *pretend* when we read the story or discuss its contents, etc.

The hard part involves talk from real or *external* contexts, when we talk about Karenina as being *a famous fictional character, well-developed and psychologically realistic, created by Tolstoy, an influence on future literary history,* etc. For claims of those sorts, though central to claims in literary criticism and history, cannot be paraphrased as talking about what is true *according to the story,* or as what the story invites us to pretend. It is this kind of external, critical discourse that brings about the need to talk about fictional characters *as characters*—which (I have argued (1999)) are best thought of as abstract artifacts. Stephen Schiffer (1996, 158–159) brings out a similar point nicely in imagining a society that tells stories, but has no terms for fictional characters (1996, 158–159). As I would now put it, terms for fictional characters are not essential for writing or appreciating literature. But they do function to *enable us* to engage in certain kinds of discussion in literary criticism and history—when we wish to discuss the origins, sources, and influences on and of various characters, the ways they are developed, their impact on later literary history, etc.

Do we have similar needs in engaging in cinematic history and criticism? Clearly we do. Those who analyze films and their history may say things like that (with apologies for bringing down the tone with these examples) Donald Duck has appeared in 197 different Disney movies (giving him the highest count of all characters), that a character that originated in one film figured as a 'crossover' character in a later film, that Simba in "The Lion King" was inspired by Kimba in "Kimba the White Lion", that the Mia Wallace character was modeled after the actress Anna Karina, that the James Bond character has been portrayed by seven different actors and appeared in twenty-seven different films, and so on. We can make sense of this kind of discourse as talking about abstract artifacts—but not as reporting what is true *according to the story* or the pretense it enjoins us to engage in. Nor are these captured in describing ways in which we engage in perceptual imagination in watching the film as we observe the images on the screen. Anyone who wishes to remain agnostic about whether to say there are fictional characters portrayed in film owes some account not just of our experience in viewing the films, but also of how to make sense of these sorts of 'external' discourse common among film critics, historians, and fans.

I am inclined to think that we *do* need to allow that we can refer to the fictional characters of films as well as those of literature, and for similar reasons: to give a proper account of external critical and historical discourse. Does this require getting into some 'substantive metaphysics' that an empiricist would do better by leaving behind? Obviously, a lot will hinge here on what one means by 'empiricism', and what one means by 'substantive metaphysics'. I was not explicit about views in metaontology when I wrote *Fiction and Metaphysics* (1999), but I have later come to develop explicitly deflationary metaontological views (e.g. in my 2015, 2020a, 2020b, 2020c) that are cohesive with what I was doing in my first book. I do not think that it's a bit of 'substantive metaphysics' to say that there are fictional characters. Instead, that there are fictional characters simply falls out of understanding how terms such as 'fictional character' are introduced—with rules that entitle us to make trivial inferences from "Jane Austen wrote a novel pretending to refer to a person called 'Emma Woodhouse' and attributing properties to her", to "The fictional character Emma Woodhouse was created by Jane Austen", to "There is a fictional character" (cf. my 2003a, 2003b). So, given the empirical fact that Jane Austen did so write, we are entitled to conclude that there is such a fictional character. (In my (2015) terms, this is an 'easy ontological inference').

In fact, I think that a thoroughgoing empiricism that also draws on empirical work on *language,* understanding its functional roles and entry rules, will give us support for thinking that agnosticism is not the appropriate response. Instead, given the rules that entitle us to introduce terms for fictional characters (and thereby enable us to engage in external discourse), we have sufficient reason to conclude that there are fictional characters, in the only sense that the relevant term ('fictional character') has. And the same would go, with slightly different entry rules, for the fictional characters portrayed in films.

11.3 Garcia-Carril on the Metaontology of Art

My work on fiction broadened naturally into work on the ontology of art generally, and Nemesio Garcia-Carril Puy gives me the opportunity to revisit that work. Garcia-Carril addresses my view about how we can and should do work in the ontology of art—which, for me, was the first step in developing and presenting my metaontological views. He argues against my approach, claiming that it reduces the ontology of art to 'triviality', and doesn't give ontological accounts a 'substantive role in the enterprise of understanding art objects and artistic practices' (2). What is that desired 'substantive role'? Perhaps among other things, Garcia-Carril seems to hold that work on the ontology of art should enable us to cope with difficult cases, borderline cases, and cases of conflict--cases where our tacit ontological conception is 'ambiguous or confusing' (11). But he complains that I give no methodological tools for addressing those cases. As he writes: "Thomasson's methodology does not provide tools to validate and explain this substantive contribution that ontology can and, in fact, makes" (17).

The central point I want to make in response is that, actually, my approach *does* provide these tools, as one can see by reading the wider context of my work. To see this, it may help readers to put my early work in the metaontology of art back in the big picture of the neo-Carnapian meta-metaphysical position I have been working to develop over many years—of which work in the metaontology of art was the first visible step.

The point of my early work in the ontology of art was to argue that (then-) popular work done on the ontology of art goes wrong if we think of it on the model of scientific 'discovery', where ontologists purport to (like scientists) make surprising 'discoveries' about what works of art 'really are', and where 'theories' are to be evaluated on grounds of their 'theoretic virtues'. This approach often leads to metaphysical posturing, with philosophers claiming to know better than the 'folk' or practitioners in the art world, and to be in a position to completely overturn their views. Yet at the same time, such an approach to the ontology of art lacks the credentials to make such claims—indeed it has a completely mysterious epistemology—as it is entirely unclear how such 'discoveries' could be made, or how we could adjudicate among competing views. Where criteria for supporting these revisionary ontological views are given at all, they are typically quasi-scientific criteria suitable for 'theoretic choice' in the sciences, or are based on appeal to unjustified 'general metaphysical principles'.[7] But as I have long been arguing, such criteria are completely inappropriate for work in ontology, which lacks the empirical support of scientific theories, and which involves adjudicating among competing 'theories' that are nearly always empirically equivalent.[8]

My criticisms of the 'discovery' model and my defense of a positive model for answering questions in the ontology of art were both based on broad views about how we should understand existence questions and modal questions in metaphysics—views that I have refined and developed over the years since.[9] Garcia-Carril suggests that my support for the approach (which he calls 'descriptivism' but I call 'deflationism') relies solely on what he calls an 'arbitrarily chosen' view of reference-fixing (17). This is not so—I make use of a much wider range of arguments, throughout my metaontological work (see my 2015, 2020a, and elsewhere). I do appeal to a

[7] Occasionally, arguments are made instead based on alleged contradictions in the relevant concepts, or poor fit with our ordinary practices. Such arguments must be treated on a case-by-case basis.

[8] For a more thorough discussion of why the criteria for scientific theory choice are inappropriate in metaphysics, see my (2025b).

[9] The most developed version of my approach to existence questions is in *Ontology Made Easy* (2015). The fullest development of my views about metaphysical modal claims is in *Norms and Necessity* (2020a), though see also the update, making use of work in systemic functional linguistics (2023b).

view of reference-fixing in support of my view in my early papers on the ontology of art—but this is not an 'arbitrarily chosen' view. Instead, I argue that direct reference theories face a 'qua-problem': there are always many ontologically/modally distinct things we could *aim* to refer to in grounding a term's reference (say, we could aim to refer to a statue or a lump or a temporal part of either; in the case of a symphony we could aim to refer to a performance, a pure sound structure, an indicated structure, etc.). Disambiguating among them (to determine whether and if so to what a term refers) requires that would-be grounders have some background ontological conception of what sort of thing the term is to refer to, if it refers at all.

Garcia-Carril rejects my view that reference-fixing requires ontological disambiguation, by citing Dodd's (2012) appeal to Gareth Evans' (1973) view that referential ambiguity may be dispelled by determining which thing (or kind of thing) is the 'dominant or causal source of the relevant body of information' associated with the term by those who ground its reference (10). But as I argue more thoroughly elsewhere (2025a), Evans' suggestion was not designed to disambiguate among purely *ontological* alternatives—which typically differ *not* in their causal powers, but only in their modal properties. It was designed to disambiguate, say, which *person* a certain name refers to—where distinct people are distinct causal sources of information. By contrast, there is simply *no prospect* of determining among merely *ontological alternatives* which is the 'dominant' source of information about *The Burghers of Calais* or *Beethoven's Ninth Symphony*. I urge readers interested in this issue to read my full reply there (2025a) before accepting Garcia-Carril's claim that this approach to reference fixing is 'arbitrarily chosen'. And I humbly ask readers interested generally in the prospects for a deflationary versus quasi-scientific metaontological approach to consider the far wider range of arguments I give for my deflationary metaontological approach in my (2015, 2020a, 2025b) and elsewhere, which there is not space to repeat here.[10]

At any rate, let us return to consider this broader metaontological view. The underlying point of my work on the metaontology of art was to insist that such facts as there are about the existence conditions, identity conditions, and persistence conditions of works of art are not quasi-scientific, discoverable features of the world, but instead are "fixed by human intentions and practices" (see my 2010, 128). As a consequence, these conditions may have places of ambiguity, open-endedness, and conflict, so that the corresponding facts are bound to have places of indeterminacy, and answers to the ontological questions are "only answerable to the extent that the way the reference of the term is grounded (and regrounded) disambiguates the ontology by determining the general kind or category of entity that the term is to refer to, if it refers at all" (Thomasson, 2010, 120). There may be many different ontological kinds of 'art', and new kinds may be introduced (Thomasson, 2010, 121). Where there are gaps and indeterminacies, we must make *decisions* about how we should

[10] Garcia-Carril also refers to a paper by Cumpa and Bueno (2021) arguing that even if we can get easy ontological inferences to the existence of works of art, the view remains 'neutral' about their natures. My arguments about how the ontological 'natures' of things are to be determined are laid out in my views on modality (2020a). Please also see my direct reply to Bueno and Cumpa in my (2020c).

talk and direct our practices, rather than making *discoveries* of covert ontological facts (see, e.g., my 2010, 120, 124). It is also this that ensures that the relevant ontological conceptions are open to *revision,* and so that work in the ontology of art is relevant to our ongoing, and constantly developing, art-world practices.

My underlying metaontological approach is a neo-Carnapian one.[11] Carnap (1950/1956) distinguished between *internal* and *external* questions. Roughly, *internal* questions about, say, the existence of an entity of a given kind (propositions, numbers, symphonies, paintings) are questions to be answered *using* the relevant conceptual or linguistic framework. These were the questions about the ontology of art I was predominantly focused on in my early (2004, 2005, 2006) papers on the topic. But we can also ask what sound like 'metaphysical' questions in an *external* mode, asking instead *what linguistic or conceptual framework we ought to adopt* (for various purposes). My more recent work on metalinguistic negotiation (2016) and conceptual engineering (2020b, 2021a, 2021b, 2022) has focused on the *external* project. Work on ontology of art is not confined to what Garcia-Carril (following Dodd) calls the 'descriptive' task of determining how our art concepts *do* work, and what tacit ontological conception that reveals. It also may include (and at its most interesting ends typically does include) work on what linguistic and conceptual (and correspondingly: ontological) conceptions of art we *should* adopt—whether to better understand or improve our interpretive, conservation, display, or other art-world practices, to keep pace with developments in the art-world, or for other purposes.[12]

Garcia-Carril argues that my approach can't meet the goal he sets for the ontology of art, "to increase our understanding of artworks and the practices to produce and properly appreciate them" (8), and that it can't help us cope with difficult cases, borderline cases, cases of conflict; cases where our tacit ontological conception is "ambiguous or confusing" (11). But that is because he only considers half of the story—the internal, not external, side of the story.

I actually agree that work in the ontology of art can and should help us work out how to deal with cases where our tacit ontological conception is incomplete, ambiguous, confusing, outdated, inappropriate, or in need of changes. But the right approach isn't to pretend we are scientists and making deep 'discoveries' about the 'real (ontological) facts'. Instead, I have tried to show (2020b) how we can develop a *pragmatic* approach to conceptual engineering to help us navigate these normative questions about what linguistic and conceptual scheme to choose, and how we might need to modify it as our needs, the world, and our understanding of the world change. In this way, the total deflationary approach to metaphysics I have developed over the past two decades can help us better deal with these questions, too—and build an approach to the ontology of art that is relevant to our art-world practices, while avoiding epistemological mystery.

---

[11] The connections to Carnap are most explicitly drawn out in my (2015, Chap. 1).

[12] And in this way, the approach harmonizes nicely with parts of Davies' (2004) work.

11.4 Donato-Rodriguez and Falguera on Abstract Artifacts and Theoretical Entities

As mentioned above, I develop the idea of 'abstract artifacts' in characterizing the ontology of fictional characters, as well as many other abstract social and cultural entities such as works of literature, symphonies and laws of state. Xavier de Donato-Rodriguez and José L. Falguera develop an interesting application of the notion of 'abstract artifacts' to the theoretical entities posited in natural scientific theories. I thank them for the suggestion—it is certainly an interesting idea worth exploring further. Here I will raise a couple of questions I have about how this idea would work, as well as addressing a couple of interpretive issues about my work and Carnap's.

First, I would like to clarify a couple of interpretive issues. The authors claim that simple realism and metaontological deflationism are independent, in part because 'from a strictly Carnapian view' one could be a metaontological deflationist and not accept simple realism, understood as realism 'in an external sense'. To clarify, I do *not* mean to be embracing realism 'in an external sense'—whatever that would mean. On the interpretation of Carnap I develop in Chap. 1 of *Ontology Made Easy* (2015),[13] internal questions are those asked *using* the terms in question, with their rules of use established by the relevant linguistic framework ('number', 'property', 'table'). These are the existence questions that generally receive a positive answer, given the easy approach to ontology, leading to a 'simple' (not explanatory) realism about numbers, properties, ordinary objects, etc. In fact, the whole of *Ontology Made Easy* may be seen as concerned with existence questions asked 'internally', which (I argue, following Carnap) is the only sense that has sense. I argue there that 'external' questions, taken straightforwardly, would have to sever the terms from the rules of use that give them meaning—and so would be meaningless pseudo-questions (see my 2015, 38–41). So I should not be taken there or elsewhere to be endorsing answers to the 'external' questions of 'hard metaphysics', nor do I think of the entities described by the sciences as "real entities in the sense of hard metaphysics" (14). Whether we are speaking of mortgages, symphonies, tables, genes, tigers or electrons, the approach to answering existence questions requires determining the rules of use for the relevant terms, including their introduction rules.[14] Sometimes the introduction rules can enable us to make easy (conceptual) inferences to the existence of the relevant entities, without relying on any empirical facts. For example, from *either* 'there are two cups on the table' or its negation, we can infer that there is a number (two)

---

[13] Following Huw Price's (2009) reading of Carnap.

[14] My remarks on natural kind terms (which I mean to include such paradigmatic examples as 'water', 'tiger', and 'gold') are not meant to suggest some kind of 'external realism' for these entities, but only to acknowledge that the rules of use for some terms (such as these) may include application conditions that are deferential to worldly conditions that will require empirical discovery (e.g. 'what is the chemical composition of gold?'). For further details see Chap. 4 of my (2020a). To acknowledge variations in the types of rules of use governing different sorts of terms is not to say that existence questions may be meaningful and answerable 'externally'—that is, while severing the terms from their rules of use.

which numbers (or fails to number) the cups. Other times ('there are tables' or 'there are Ivory-billed woodpeckers'), the introduction rules for the relevant terms include application conditions, and we need empirical investigation to determine whether application conditions are met, and so whether the existence claims are true. But in neither case do we need something beyond conceptual and empirical work—we need nothing 'epistemically metaphysical' (to use Sider's phrase (2011, 187)). And (as I argue extensively) in neither case are the methods of neo-Quinean 'hard ontology' appropriate.

One can of course instead treat external questions as (external) *practical* questions about whether or how we should use (parts of) a given linguistic framework. So understood, they may be perfectly meaningful, practical questions (2015, 40–41). Donato-Rodriguez and Falguera mention that Carnap paves the way to reconsider ontological disputes as 'pragmatic matters' (14) where, by contrast, I make it 'too easy'. But throughout my recent work I have embraced and developed the idea that we can reconstrue and address many debates in metaphysics as practical disputes to be addressed by conceptual engineering (see my 2017, 2020b, 2021a, 2021b). So here, too, my view is *Carnapian,* not in tension with Carnap. The view they outline as 'even more Carnapian' than mine (20) is essentially mine—so I think we are actually on the same page here.

Let us move on to Donato-Rodriguez and Falguera's positive suggestion: that theoretical entities can be understood as abstract artifacts, in the sense in which I claim that fictional characters (and many other social/cultural entities, such as novels, symphonies, and laws of state) are abstract artifacts. I think this is an interesting suggestion, though I also have some hesitation, and some questions.

We can certainly think of the theoretic entities of the sciences in this way, and in some contexts (for example in discussions in the history of science, or in discussing the contents of scientific theories) it may be useful to do so. That is, we can think of scientific theories as 'texts', and the entities described in them as like 'characters' in works of literature. We can similarly think of the people described in texts (whether they are historical works or fictional works, characters originating in that work or purporting to be some historical figure) as fictional characters—and in this regard it may be useful to be able to speak about the features of Aaron Burr *as a character in 'Hamilton'* (or Aaron Burr as represented in David Stewart's biography *Fallen Emperor),* separately from speaking of the features of Aaron Burr, the historical man. So similarly, we may often want or need to speak of the features of the entities represented in scientific theories *as they are there represented*—we can speak (for example) of atoms, *as they are represented in X theory* distinctly from atoms *as they are represented in Y theory.*

But there are also important disanalogies between the functions and rules that govern use of fictional terms, versus of theoretical terms. Terms used in works of fiction typically have two uses: In works of fiction, authors typically use ordinary terms such as 'dog', 'tree', or 'person', for purposes of telling a story, and as readers we may engage in something like a pretense that the relevant conditions (established in ordinary speech) are fulfilled, even if we know that they aren't (we all know that there is no such person, say, as Sherlock Holmes, but we may still pretendingly use

'Holmes' as a person name). We may also use fictional names in discussions of literary history (say) to refer back to the abstract artifact created by an author, at a particular point in literary history, developed in different ways by different authors, etc. Given these two different uses, we can also (as I discuss elsewhere (2010)) distinguish two sorts of existence claims, depending on whether the assumption is that the name is used with the application conditions for a person (say), or with those for a fictional character. The claims of the first sort are false; claims of the second sort are generally true (though they can also be false, say, if we attempt to speak of the fictional character *Harry Potter's little sister,* when there is no such character).

Terms for theoretical entities introduced in scientific theories, however, can't be thought to be introduced with the same functions and rules of use as terms for fictional characters. Theoretical terms (unlike general terms used in stories for persons, trees, or dogs) often do not have an antecedent meaning given outside the theory. It is common to think of terms for theoretical entities like electrons, black holes, gravitational forces, or genes as being given meaning by the theories that introduce them. If so, it is these theories that would introduce the application conditions (or other rules of use) for the relevant terms. In short, the authors of scientific theories (unlike the authors of fiction) are not engaging in a pretense that piggybacks on the rules of use for the terms *outside the* pretense. Moreover, of course, theoretic terms are not introduced for purposes of story-telling, but rather are typically introduced as terms for things that are supposed to play a role in *explaining* our observations (even if we can't observe electrons, we can 'posit' them to causally explain traces we can observe on a screen). In short, the interest we typically take in theoretic entities is not as characters in a story (though we certainly can do this, if our interest is in the history of science), but as posited *explainers*.

Of course, abstract artifacts can't fulfill that role. Given their abstractness, abstract artifacts can't cause a trace to appear on a screen, cause light to bend, etc. Similarly, some things are literally true of abstract artifacts—as Donato-Rodriguez and Falguera put it, such a "kind of abstract entity… depends on particular concrete entities, such as the scientists that formulate the scientific theories in which the terms for these entities occur" (17). But few would want to say that it's literally true that electrons depend on, and only come into existence through, the activities of scientists. So abstract artifacts *can't* be the intended referents of the theoretic terms of the sciences, *as they are used* in those theories—they can't fulfill the application conditions that give the terms meaning in the relevant theories. We can certainly speak of theoretic entities as the 'characters' of scientific theories, and of what they are like, *according to those theories*—and if our interests lie in the history of science, it may be useful to do so. But it seems more problematic to allow that terms for theoretic entities, *as they are used in those theories*, refer to abstract artifacts.

Similar issues arise for discourse about fictional characters, of course. We have to distinguish talk about what Sherlock Holmes is like *according to the story* (a man, a detective, born to a human mother) versus what the character is *really* like (an abstract artifact created by Arthur Conan Doyle). And we have to distinguish talk about what Sherlock Holmes does and causes *according to the story* from out and out talk about the role of the *character* (the abstract artifact) in literary history. And (as mentioned above) I came to think (by my (2003a)) that we also shouldn't take the names of fictional characters *as used within a story* to refer to abstract artifacts, though their use in stories may *entitle* us to refer to the relevant abstract artifacts in our literary critical discussions. Similarly, we can certainly take terms for theoretic entities *as used within a theory* to entitle us to refer to the relevant abstract artifacts in our discussions in the history of science. But the question will remain of how we should understand theoretic terms to work *within a theory*—whether we should think they refer or not, and if so, to what. We can rest content with the idea that fictional names as used *within a story* are used merely pretensefully, without any aim of referring to real people, dogs, or trees. But the same sort of account doesn't transfer easily to the uses of theoretic terms *within a scientific theory,* which seem to be introduced with very different aims. In short, we can ask whether phlogiston exists *as an abstract artifact characterized in a theory;* but we can also ask whether *anything matching the theory's descriptions* exists and can explain what we observe during combustion. The latter seems like a remaining question of interest, even if we agree that we *can* speak of theoretic entities as abstract artifacts, when we are writing the history of science. Nonetheless, I agree with Donato-Rodriguez and Falguera that the notion of 'abstract artifacts' may have wide application to other areas of philosophical interest—including to the history and philosophy of science.

11.5 Rychter on an Easy Ontology of Facts

The neo-Carnapian approach I take to existence questions leads to 'easy' answers to many existence questions, including questions about the existence of facts. In "Fictionalism and the easy approach to the existence of facts", Pablo Rychter challenges the idea that we should accept an 'easy ontological' approach to the existence of facts. He argues instead in favor of a fictionalist approach to talk about facts, and claims to answer my challenge of saying *what more it would take* for the claim that 'there are facts' to *really* be true, as opposed to being merely *pretended.* I will address those challenges here.

On the easy approach to ontology that I develop and defend in *Ontology Made Easy* (2015), we can often answer existence questions by way of trivial inferences from undisputed premises. In the case of facts, plausibly, this would involve, as Rychter suggests, an inference such as:

1. Socrates is wise.
2. If Socrates is wise, then it is a fact that Socrates is wise.[15]
3. There is a fact (namely, that Socrates is wise).

[15] To avoid worries that 'wisdom' is unobservable or a vague term, I prefer to use more down to earth examples, such as 'the fact that the barn is red' or 'the fact that this electron has a negative charge'. But for coherence, I will follow through with Rychter's example here—if the term 'wisdom' concerns you, please just shift to a different example.

Such inferences entitle us to accept that there are facts, where this is a form of what I call 'simple realism' that (unlike David Armstrong's explanatory realism) doesn't make or rely on any *explanatory claims* (for example, that facts are needed to 'explain' how truth is grounded).

Premises such as (2) above are based on a conceptual truth, e.g. that if an object o has a property P, then the fact that that o has that P obtains.[16] And this conceptual truth, I hold, reflects a rule of use for introducing 'fact' talk: that in ordinary English we are, without any further evidence, entitled to make the inference from 'Socrates is wise' to 'it is a fact that Socrates is wise', and so on for other introductions of 'fact' talk. Moreover, given these rules (I argued (2015)), the fictionalist about facts has no way of articulating *what more it would take* for claims about facts to 'really' be true, rather than being mere pretenses.

Rychter's resistance to the easy argument comes in here—for he denies that there is such a conceptual truth. For, as he writes, "it is *conceivable* that *o* is *P* and 'fact' fails to apply because there is nothing, other than o or *P*, for 'fact' to latch on to." (7) Why? What more might one think is required for the fact that Socrates is wise to obtain (what could we expect to 'latch on to')? According to Rychter, what more is required is that "there *exists an extra thing, x*, that is not identical to any of its components" (8, italics mine). And by appealing to this, Rychter also claims to respond to my challenge to fictionalism—to articulate what more it would take for claims about facts to 'really be true' rather than just pretended.

As Rychter notes, the thought that what more is required is that *'there exists an extra thing'* is a line of thought I argued against extensively in my earlier work (especially 2007a). Cohesively with the easy approach to ontology, to ask whether there is 'an extra thing', we must ask for the application conditions for 'thing': what would it take for there to be some 'composed unified structure', or some 'extra thing'? I identified there three uses of 'thing': (1) as a term for a 'Spelke-object' (a roughly unified, trackable, independently mobile lump of stuff); (2) as a 'covering term', such that 'thing' is guaranteed to apply if any first-order sortal term applies (if there is a ball, or a protest, or a speck of dust, or…); and (3) the alleged 'neutral' use of heavyweight ontologists. I have seen no clear sense made of the application conditions for, or other coherent rules of use for, the heavyweight ontologist's use of 'thing', and I think that we have every reason to suspect that this term is not being used in a way that gives us fully meaningful, truth-evaluable claims at all—and that this is part of what makes the heavyweight ontologist's disputes unresolvable. While Rychter seems sympathetic with the heavyweight ontologist's use (see p. 11 n. 3), he does not try to defend it here. That leaves us with (1) or (2). Rychter agrees that (1) can't be what's relevant here (anyone who thought that 'the fact that the barn is red' should refer to a separate additional lump of stuff (beyond the barn) is essentially making a category mistake and failing to understand how 'fact' talk works). That leaves us with (2).

[16] For simplicity, I will leave out the relational case here—take it as implied throughout.

This is where Rychter claims to give a response to my earlier arguments. What more is required, he claims, for 'It is a fact that Socrates is wise' to be *really* true, is "the existence of an *object* constituted by Socrates and the property of being wise, and not to be identified with either of those other two things" (17), where 'object' is used in the *covering sense.*

But I cannot see how this helps. For, in the covering sense, 'object' is just used as a term of generalization—a place-holder for genuine sortal terms. And it is guaranteed to apply if the application conditions for any (genuine) sortal term are fulfilled. The most relevant sortal, in this case, obviously is 'fact', not 'building' or 'squirrel', or.... So, what Rychter has given us, in response to my question, is essentially this. I ask: what more could it take for 'fact' to apply, other than that some object instantiate some property? He replies: What more it could take, is for there to be some other 'object' (*in the covering sense*—where 'fact' is among the terms covered). That is, in response to my question: 'What more would it take for 'fact' to apply?', we get only the answer: For 'fact' to apply. But that is no answer, it is just an empty triviality. Moreover, if he is assuming that for 'fact' to apply would require that *more* conditions be fulfilled than just for an object to instantiate a property, then he hasn't yet said *what* more is required, and so hasn't addressed the challenge to fictionalism.

Now some might be inclined to think that we just reach a standstill here—are the application conditions for 'fact' just that some object instantiates some property (as I hold), or is there (as Rychter holds) more required—that there be some other 'object' there than just Socrates and the wisdom? But I don't think this is the right reading of the situation. First, Rychter has given us *no clear content to the claim at all—no sense of what more is required.* It is not clear that he is meaningfully articulating different application conditions at all.

By contrast, my view not only articulates application conditions (or perhaps better: rules for introducing fact talk) clearly, these are also rules that reflect the ordinary use of 'fact' in English. For in ordinary English conversation, we are entitled to introduce 'fact' talk without requiring any additional evidence. Imagine the following courtroom dialogue:

- Prosecutor: It's a fact that the gun was found in the defendant's car.
- Defense attorney: No it isn't.
- Prosecutor: Are you denying that the gun was found in the defendant's car?
- Defense attorney: No: the gun was found in the defendant's car, but it's not a fact that the gun was found in the defendant's car. There are no facts.
- Everyone else in the courtroom: *Whaaaaat?*

There is something Moore-paradoxical about the defense attorney's second assertion: The gun was found in the defendant's car, *but it's not a fact* (that the gun was found in the defendant's car). In ordinary English, it is hard to see the defense attorney as making any comprehensible speech act (they seem to be both asserting and taking back the assertion). ("But it does make sense *in the ontology room*", some will want to say. But, as I have argued, what we have in the 'ontology room' is merely an *appearance of sense* by introducing a use of 'object' lacking in meaning, along with adopting inappropriate criteria for resolving what are thought of as 'ontological debates'—criteria I have argued against extensively elsewhere (2007a, 2015, 2025b).)

I argued in *Ontology Made Easy* that the ordinary rules governing fact talk entitle us to make easy inferences to the existence of facts—it is redundant to say, "The gun was in the defendant's car, *and* it is a fact that the gun was in the defendant's car". But (one might ask) why would we want to add 'fact' talk to our language, via *redundant* inferences that require no additional 'evidence'? I can now answer that question, based on work from systemic functional linguistics.[17] Introducing talk of 'facts' is a way of introducing a new noun term from a proposition—so from 'The barn is red', we may refer to 'The fact that the barn is red'. This involves introducing what Michael Halliday (2009) calls 'grammatical metaphors'—shifts in grammatical category. As Halliday shows, grammatical metaphors are useful in a language because they enable us to do new things with language—including showing relations among ideas across a longer text, and generalizing in ways important to reasoning, and to developing scientific theories and bureaucracies. In the case of 'fact' talk, we can easily see the use of introducing reference to 'facts', not to 'pick out' or 'track' some 'additional object' in the region, but rather as a way of enabling us to generalize and lay out general principles of reasoning. For example, it may be redundant to simply move from 'The gun was in his car' to 'It is a fact that the gun was in his car'. But introducing talk of facts enables us to generalize and say things like 'Everything the prosecutor asserted is a fact', or 'There are facts that do not fit with the defendant's account of the case'. It also enables us to lay out general principles of legal reasoning, such as: 'If there are facts that don't fit a defendant's account, then that account should be rejected', or 'in a jury trial, the jury's job is to decide the facts'.

If this is the correct account from linguistics of the functions of fact talk, and of the redundant inferences by which that talk is introduced, then we *do* have reason to accept that it is a conceptual truth that *nothing more is required* for the fact that o is P to obtain, than that o is P. There is no reason to go fictionalist, rather than simple realist, about fact talk, and fictionalists remain with the unanswered challenge of saying what more we should think is required for there to *really be* facts, and *why* we should think these additional demands hold.[18]

[17] See, e.g. Halliday (2009). For summary of some of this work, see my (2022). I had not yet discovered the work in systemic functional linguistics when I wrote *Ontology Made Easy.*

[18] The same form of argument applies not just for talk of 'facts', but also of 'propositions', 'properties', 'events', and many other sorts of terms that serve as grammatical metaphors, introduced via redundant inferences to serve textual functions.

11.6 Sambrotta on Ontological Discourse

Mirco Sambrotta's paper brings together my work on easy ontology with some of the neo-pragmatist themes of my most recent work. He argues that my account in *Ontology Made Easy* doesn't do sufficient justice to ontological discourse and what it can do for us, and he raises a good question: Why do we do ontology? What do ontological vocabulary and constructions do for us? (8–9).[19] He also gives a pragmatic answer that I am very sympathetic to.

For me, there are (at least) two different questions we might raise about 'ontological discourse', depending on what is meant by 'ontology' and 'ontological vocabulary': One is, why does our shared public language include terms such as 'exist' or 'there are'? What are the functions of such terms? The other is: what is the use of doing *philosophical* work in ontology, of debating about what our 'ontological commitments' are, about whether there *'really are'* tables, numbers, or trees, and the like? I am happy to assume that the former has some interesting positive answer—that there are functions such terms serve in our shared vocabulary. But I would not simply assume that all of the discourse of philosophical ontology (that asks about 'ontological commitments' or what there '*really* is') makes sense and serves some useful functions. So I would not go so far as to join with Sambrotta in taking for granted that "ontological practice is itself a legitimate discursive social practice" (9). There may be problems, at least with some parts of contemporary metaphysics, as I have tried to show throughout my work.

So, let us turn to the uncontroversially good first question: What are the functions of having a term like 'exists' in our language? In answering this question, Sambrotta offers various suggestions about the functions of 'ontological judgments': Ontological judgments "determine what should and should not count as evidence in support of them" (10), express "norms of reasoning", endorse "a certain epistemic principle", "articulate what should or should not count as a factual claim", elucidate "norms governing our social practice" (11), "express commitments to the adoption of [linguistic or conceptual] frameworks" (11), etc. To my ears, these suggestions seem potentially different from each other—so I would want to move more slowly—asking which sorts of 'ontological judgments' are at issue in each case, and which of

[19] A few clarifications are also in order. I do not (as Sambrotta says) see "all our existential claims as genuinely descriptive" (4) (though much depends on what one means by 'descriptive' here). I do think that such claims, interpreted *internally,* generally can 'easily' (through empirical and/or conceptual means) be assessed as true or false; but I deny that they aim to describe some property of existence possessed by an object tracked in the world. I also don't use terms such as 'substantial' or 'robust' realism, factuality, etc., as I find them typically unclear. I also do not hold that *ontological conclusions* play no 'explanatory or justificatory role'; I simply hold that *numbers, propositions, properties,* and other entities the easy ontologist accepts aren't to be treated as 'posits' to 'explain' anything, as I reject the explanatory power approach to answering questions about what exists (see my 2025b). It is also worth noting (p. 9 footnote 26) that one can find more than a 'trace' of distinctions between different functional roles of different parts of language in any of my recent work on modality, conceptual engineering, and linguistic functions (see, e.g., 2020a, 2022, 2023b, 2025b).

these functions each type of 'ontological discourse' is supposed to fulfill.[20] But some of the answers Sambrotta gives I not only like, but explicitly endorse. For example, he argues (in a Carnapian spirit) that ontological discourse expresses "commitments to the adoption of conceptual structures within which alone description is possible" (13–14). I develop an idea along these lines explicitly in my (2017) paper on metalinguistic negotiation, arguing that "some classic philosophical debates about existence can be readily seen as reflecting disagreements… about whether the term *should be used at all*" (2017, 16).

Sambrotta briefly notes the parallels between my view there and the view he defends here (13). But he expresses puzzlement about how this understanding of ontological debates fits with my easy approach to ontology (2015). As he writes:

> …it is unclear how to reconcile this view [that ontological disputes can sometimes be understood as implicitly advocating for views about whether certain terms or concepts should be used] with [the] metaontological deflationism (and first-order ontological realism) [Thomasson] advances in most of her works… if ontological disputes are easy to settle, they will turn out to be factual debates of some sort rather than metalinguistic negotiations (13)

So, this provides a good opportunity for me to clarify how these parts of my work fit together. We can, following Carnap, distinguish two senses in which existence claims may be made: internal or external. As I say in my (2017), the work of *Ontology Made Easy* is about existence questions taken in what Carnap calls an 'internal' sense, which I interpret (following Price, 2009) as questions asked *using* the terms of the relevant framework (with their customary rules of use intact) (2017, 6–7). If we want to simply ask "do numbers exist?", "do tables exist?", "do properties exist", and so on, *using* these terms (both 'exist' and 'numbers', 'tables', 'properties') in their ordinary senses, with their ordinary rules of use intact, we can answer these questions easily: Yes. For we can derive claims of their existence from uncontroversial claims, in accord with the standard rules of use for these terms (see my 2015, Chaps. 2 and 3).

Nonetheless, what serious ontologists are *doing* in uttering such obvious truths or falsehoods may often be interpreted instead as addressing *external* questions, where these are taken not as (meaningless) factual questions, but rather as *practical* questions about whether we should adopt the relevant linguistic framework (see my 2017, 9). If we interpret what they are doing in this way, those who deny that tables exist may be seen as uttering a literal falsehood, but as implicitly *recommending* that we give up talk of composite material objects, e.g., since it leads us into puzzles and contradictions. Those who deny that numbers exist may be seen as recommending that we give up the use of number terminology, given, perhaps, presumed conflicts with commitment to a physicalistic or naturalistic view of the world (see my 2017, 17–18).[21]

---

[20] I would be more inclined to think (following Restall, 2005; Ripley, 2013) that it is *logical* claims that express norms of reasoning (at least where these are formal norms). Metaphysical modal claims, I claim (2020a, 2020b, 2020c) express certain kinds of *semantic norms* and what follows from them (and these may be treated as a certain kind of norm governing a social practice of speaking).

[21] It is worth noting, however, that many ontologists resist this interpretation of their debates, that I am not committed to seeing *all* ontological disputes in this light, and that this move still won't make all debates, or all moves, sensible. Once we are carefully attuned to the actual functions served by our number terms, say, or even our terms for ordinary material objects, we can see pretty

This assessment of the role of *external* claims, practically interpreted, fits well with views I have been developing more broadly elsewhere: that much of traditional philosophical work, including in metaphysics, can be helpfully reconceived as work on the *practical* questions of what linguistic or conceptual scheme we *should* use, and how we should use it (see my 2017a, 2017b, 2020b, 2021a, 2021b, 2025b). One point of difference from what Sambrotta says is worth bringing out here, however. That is: I would *not* say that debates about such questions "lie beyond the scope of rational resolution" or "cannot be in the market of rational considerations" (10). On the contrary, I have argued (2020b) that many of these debates may be addressed in pragmatic terms, provided we are clear about what functions the relevant parts of language are to serve.

Once we see the difference between internal and external readings of existence claims, we can see how to reconcile these two sides of my work. (And the distinction between internal and external claims lines up with different things we can be *doing:* asking (first-order) 'factual' questions while *using* our extant language or concepts; or addressing metalinguistic questions about what linguistic or conceptual scheme we *should* be using). As a result, I don't see Sambrotta's view as telling *against* any part of my view. Instead, I see him as finding his own way to very similar conclusions to those I develop and discuss at length in my post-2015 work, and developing them in an interesting way. It is always heartening to find others driven to the same sorts of conclusions.

11.7 Dohrn on Modal Metaphysics

While I develop a deflationary approach to existence claims in *Ontology Made Easy,* I develop a deflationary approach to metaphysical modal claims in *Norms and Necessity.* Daniel Dohrn addresses my approach to metaphysical modality, arguing that it unduly rules out 'substantial metaphysical inquiry' about certain deep metaphysical modal claims—for example, about personal identity and fundamentality.

Before evaluating how normativism fares on these issues, however, it's worth pausing to correct some misunderstandings about my view. (Readers interested in my views about metaphysical modality are referred to my (2020a) book *Norms and Necessity*. Curiously, Dohrn does not discuss or cite this work, which gives by far the fullest development and defense of my views on modality—the views he aims to criticize.)

The first point to clarify concerns the subject-matter of metaphysical modal claims, and the second concerns their mind-independence. Dohrn alleges that modal normativism "chang[es] the subject matter" of modal knowledge (1) as it "comes with a hypothesis about the subject matter of modal claims… they state semantic

clearly that it would be a terrible idea to remove such terms from our vocabulary. For further discussion, see my (2025b, Chap. 9).

rules" (5). Against this, he argues that it "violates the intuition that the subject matter of a claim should be determined independently of asking whether and how we know that claim" (8). But this is a misunderstanding. I do not say that metaphysical modal claims are *about* semantic rules nor that they *state* semantic rules—modal normativism is not a claim about the *subject matter* of metaphysical modal claims. Instead, I take pains to distinguish my normativist view from classic conventionalist views (as standardly—and uncharitably—interpreted) (see my 2020a, 27–31)—and explicitly deny that on my view, metaphysical modal truths (such as 'the statue could not survive drastic changes of shape') are *about* our linguistic conventions or semantic rules or *made true by them.* The central thesis of normativism is that modal metaphysical language doesn't have the function of *tracking* or *describing* at all—neither possible worlds, nor modal features of this world, nor our linguistic conventions. Instead, the function of metaphysical modal vocabulary is to enable speakers to convey, enforce, and re-negotiate semantic rules in useful ways, while remaining in the object language (Thomasson, 2020a, 58). But, given easy ontological inferences, we can still take modal claims to be (in a harmless sense) 'about' modal facts and properties (2020a, 137–138). As I insist, "the literal semantic content *is* world-oriented content about [e.g.] statues, and may say something true about the world" (Thomasson, 2020a, 138)).

The reason this subtle point makes a difference is not just that it shows accusations about misplacing the 'subject-matter' to be baseless. It is also essential to avoid accusations that the view leaves metaphysical modal claims *mind-dependent.* This is the second (and even more important) point to clarify. Dohrn repeatedly asserts (allegedly against my view) that "It is not our practice but the independent structure of reality that fixes… modal profiles" (2, see also 20, 22, 23), and that normativism "jeopardize[s] our notion of an independent reality" (20). This would follow if one took the conventionalist view that modal claims *describe* semantic rules, so that it *requires the adoption of certain semantic rules to make them true.* But I repeatedly take pains to show that this is *not* the normativist view, and that the normativist approach does *not* make the truths of metaphysical modal claims contingent on our practices (see my 2020a, 90 and 170–171)—though of course the *meanings* of these (and all) claims depend on our linguistic practices. On the contrary, as I argue, the modal normativist is entitled to accept independence counterfactuals such as "even if there had been no language, 'Seals are necessarily mammals' would have been true" (2020a, 81–83 and 158–159). Dohrn's suggestion that we accept the normativist view for mind-dependent entities (such as artifacts and works of art), but not for 'independent' parts of the world also relies on this misunderstanding.

One final point of correction: it is not true that my only argument for modal normativism is the epistemological one.[22] In the book (2020a), I lay out separate epistemological, methodological, and ontological advantages of modal normativism. Moreover, in my most recent work (2023b), I have added to this evidence from

[22] Nor is it true that I only criticize Lewis' modal epistemology, as Dorhn suggests (11). Instead, I consider a wide variety of recent approaches (2020a, 147–185).

systemic functional linguistics that supports (and enriches) my account of the functions of modal language. In assessing the view, I ask readers to consider *all* of these reasons in its favor.

But let us now turn to the central point of Dorhn's paper: that there are certain 'substantive' metaphysical modal truths that modal normativism can't account for, and that we should not rule out. He gives three examples of cases in which serious metaphysical inquiry is called for (2), since (he alleges) appeal to semantic rules plus empirical data couldn't decide the issue:

1. "I could undergo fission and be precisely one of the two resulting persons"
2. "No particle could be both positively and negatively charged" (21)
3. "Necessarily, the whole is fundamental" (23)

Let us begin with the second. This is presumably a question of *physical* (or 'nomological', not metaphysical) possibility or impossibility. If someone were to seriously ask me, whether (2) is true, I would suggest that they ask a physicist—not a philosopher. Matters of physical possibility I don't discuss (much) in the book; though I do suggest that the view of metaphysical possibility I develop would fit well with a treatment of physical necessities along the lines of Sellars', as giving object-language expression of norms of inference based on empirical evidence—regarding which sorts of inferences succeed and which fail, and which inferential rules serve well in a predictive and explanatory theory (see my 2020a, 121–122). But it is to physics we must look to know what the empirical evidence is, and (accordingly) which inferences serve well in a predictive and explanatory theory.

The first case, concerning claims of personal identity, is more interesting. For claims about personal identity more plausibly concern claims of *metaphysical* modality that have long been of central interest in metaphysics. Dohrn suggests that a normativist could (implausibly) deny that there is an answer to this question (20), but in a footnote (5, p. 20) concedes that a normativist might say that there is an answer, but that we lack some relevant empirical data. The latter is just what I would be inclined to say. The 'I' that is central to our self-conception comes with a claim of continuity of experience *from the first-person point of view.*[23] When I wonder whether I would survive fission, or who I would be, a central condition for continuing to apply the 'I' is the continuation of that sort of experience, as is knowable only from the first person point of view (though it is still a sort of *experience,* and thus empirical). It is not a case in which we need some deeper metaphysical 'inquiry' to 'tell us' what the answer is (how, precisely, could that go?). It is a case in which the relevant modal facts can be determined by nothing more than grasping our concept of person (which includes a tie to the continuity of first-person experience) and having empirical knowledge (about how that, perhaps among other empirical facts, would go). Since no one has undergone fission, no one at this stage can say. But this is not a counterexample involving some deep 'metaphysical' fact about modality that normativism can't reach.

[23] This is brought out nicely in much of Lynne Rudder Baker's work on personhood.

What then of the third case—of Jonathan Schaffer's claim (2010, 61–64) that it is necessary that the whole is fundamental? It is worth noting that a long-term goal of my work (from *Ordinary Objects* (2007) onward) is to demystify metaphysics—not to preserve all appearances of 'deep and substantive' metaphysical debates claimed by its practitioners. As such, I have long aimed to identify and root out claims or questions that seem to make sense on the surface, but turn out to be based on misuses of language, faulty presuppositions, or other difficulties. I would not assume that any claim made by a prominent metaphysician must be worthy of serious inquiry. Without further details about how terms such as 'whole' and 'fundamental' are being used here, it is difficult to assess whether the above claim is fully meaningful and truth-evaluable, or how we could come to know whether it is true or false.[24] Further assessing this issue would require a great deal more discussion than there is space for here. I will have to leave it by saying simply that it is far from clear that we have here a substantive and interesting metaphysical modal claim that must be true or false, and yet that we cannot establish by way of some combination of semantic rules and empirical work (even Schaffer thinks that empirical matters are at least *relevant* here, and they form half of his argument; and anyone should accept that we must understand the meanings of our terms before we can assess the truth-value of claims they figure in). In any case, I thank Dohrn for raising these cases, as examining them can give us a better sense of the resources available for the normativist to approach various kinds of modal claims.

11.8 Locke on Modal Normativism and Possible Worlds

Theodore Locke's paper works to extend my modal normativist framework to forms of discourse I did not discuss much in *Norms and Necessity,* namely talk about possible worlds, 'ways things could be', and counterfactuals.[25] He develops valuable suggestions for how to extend a normativist account to these other forms of modal talk, providing interesting and helpful ways of further developing the normativist framework.

Possible worlds talk can give perhaps the strongest illusion that modal talk aims to track and describe certain entities posited in metaphysics: that it is *about* possible worlds—treated along David Lewis's (1986) lines, as 'concrete worlds' spatio-temporally and causally isolated from us—and that the latter serve as the truthmakers for modal claims. But why would we want possible worlds talk? In *Norms and*

[24] I also would reject the presupposition of Schaffer's inquiry, that there is one "metaphysically correct way to carve up the cosmos" (2010, 46).

[25] In *Norms and Necessity* I say only a little about possible worlds talk (see, 2020a, 132–137), arguing that we can accept possible worlds through 'easy' ontological arguments (2020a, 135), so that we are indeed entitled to quantify over possible worlds. I don't discuss 'ways' talk at all, and only discuss *empirically grounded* counterfactuals, not metaphysical counterfactuals (see 2020a, 121–122 and 153–154).

Necessity I suggest (following Lewis) that doing so adds expressive power to our language (2020a, 133) in ways that, for example, enable us to express philosophical theses (of global supervenience, etc.) that we could not otherwise express.

Locke goes beyond this, developing a far more detailed account of what difference it makes to add possible worlds talk to our modal vocabulary—while still remaining cohesive with the normativist idea that modal talk functions primarily to convey, enforce, or renegotiate semantic rules and their consequences (while remaining in the object language). Talk of possible worlds, on Locke's view, "provides a different syntactic framework for offering alternative illustrations of semantic rules, or consequences of those rules" (13) for "talk of worlds, ways, and possibilities is just a syntactic metaphor for talk about what is possible and necessary" (14). In saying this, Locke draws on work from systemic functional linguistics, work that is extremely helpful for justifying the claim that we can get trivial inferences to claims that there are possible worlds, and for understanding why we would want to add nominalized terms for possible worlds and the like to our modal vocabulary.[26] Locke builds on the idea from systemic functional linguistics mentioned above: that language often permits shifts in grammatical category to introduce what Michael Halliday (2009) calls 'grammatical metaphors' (Locke uses the term 'syntactic metaphors'). Introducing grammatical metaphors, as Halliday notes, is typically useful since it adds *textual* functions to a language, in ways that enable us to show connections among ideas in longer texts, and to form generalizations that enable us to build complex scientific theories and bureaucracies (2009, 138). This fits neatly with David Lewis's (1986) observation that speaking of possible worlds adds expressive power to our (philosophical) theories.

But as Locke shows, there is still more that can be said. Halliday notes that introducing nominalizations enables us to *quantify and qualify* in ways we could not without the change in grammatical category: we can say "they danced in the second act" with the verb 'danced'. But we can only say "they did three Romanian folk-dances in the second act" if we shift from the verb form 'danced' to the nominalization 'dances'. Much the same (as Locke shows) can be said about possible worlds talk, as it enables us to serve functions beyond those we can reach with simple modal verbs. While we can say, 'the Patriots could have won the game last night' with just a modal verb, possible worlds talk enables us to *quantify* various *ways things could have been different*—saying, e.g. '*there are at least three ways the Patriots could have won the game…*' It also enables us to *qualify* talk of possibilities, so that we can say, e.g. that some of those possible worlds are *more similar* to the actual world than others (these are 'more nearby' worlds, as the saying goes). None of this can be achieved with simple modal talk that merely insists that the Patriots *could have* won.

While Lewis emphasized the importance of possible worlds talk to articulating theoretic differences in philosophy, something I particularly like in Locke's paper is the way he shows that informal talk along these lines, talk of 'ways things could have been', plays important roles in everyday contexts as well. For, as he shows, it helps

[26] For an updated development of my views on modal discourse, expanded to take into account work in systemic functional linguistics, combined with other empirical work on the development of modal thought and talk, see my (2023b).

us with our 'conversational scorekeeping'. For, "talk of ways (worlds or possibilities)… [enables] speakers to change permissibility facts in a game of linguistic scorekeeping" (19–20). In everyday contexts, if a speaker says, "there is no way the Patriots could have won last night", they attempt to add to the conversational scorecard that saying the Patriots could have won is 'out of bounds'. But by describing different *ways* they could have won (e.g. by kicking a field goal in the last second), the interlocutor justifies removing this prohibition from the conversational record. Similarly, in uttering 'It's necessary that water is H_2O', chemists make any claims about water's being of some other chemical kind 'off-limits' (21–22), foreclosing 'any conceptual or empirical inquiry' into ways water could be of a different chemical kind. We can then see talk of possible worlds as a way of formalizing the useful roles that 'ways' talk has in ordinary speech, in noting or adjusting the required and permissible moves in a conversation and enabling speakers to "change permissibility facts in a game of linguistic scorekeeping" (20).

Another important contribution of Locke's paper is extending the normativist approach by showing how counterfactuals, too, may be seen as conveying rules and permissions. Claims of the form 'if P were the case, then Q would be the case', he suggests, "express a semantic requirement to accept that Q given P as a hypothetical supposition and a range of relevant auxiliary assumptions determined by the context" (25–26). I would only add this: that in some cases the relevant requirements and permissions (to make the inference under the hypothetical supposition) may be *semantic*, and in others they may be *empirically grounded*. 'If Sam were married, then she would have a spouse' may convey a consequence of semantic requirements (to accept that Sam has a spouse if we accept that Sam is married); correlatively, 'If Sam were married, then she would have a husband' may be countered by 'No, she *might* instead have a wife'. Some such counterfactuals do play roles in metaphysical arguments, e.g. 'If minds were physical objects, then they would have extension', or, 'If a work of literature were identical to a copy of it, then it could be destroyed by fire'. These I am happy to see, with Locke, as expressing semantic requirements to accept the consequent, on hypothesis of the antecedent. Other counterfactuals, e.g., 'If the match were struck, it would ignite' may convey a requirement to make an inference, not given the semantic rules alone, but given past experience (with matches or the relevant chemicals). In either case, the relevant counterfactuals can be treated as conveying inferential permissions and requirements (that apply given certain hypothetical suppositions and auxiliary assumptions), even though there is variation in which norms are at stake. As Locke aptly suggests, one way of laying out the relations between them is to say that where the relevant inferential requirements come from semantic rules alone, those inferences are "maximally robust insofar as the addition of *any* range of relevant auxiliary hypotheses will not infirm the inference between p and q" (26).

In any case, I heartily agree with Locke's chief conclusion, that we can retain talk of possible worlds without the need to 'fit' such worlds into a naturalistic ontology. The work he does in showing the functions added to language by talking of *ways things could have been*, of *possible worlds*, and of talk using counterfactuals, is an important contribution to the normativist project, and is precisely the sort of thing we need to show why such talk should be retained, and why we can use it without concerning ourselves with spurious metaphysical problems.

11.9 Olivero on Modal Normativism and Artifactual Kind Terms

Irene Olivero makes an important contribution that connects portions of my earlier and more recent work, by applying some of my (2020a) work on modal normativism to debates about the meanings of artifactual and natural kind terms. There have been extensive and high-profile debates about whether *all* kind terms work on the direct-reference model that has been so popular for natural kind terms (these were debates I entered into in some of my earlier work (2003c, 2007b)). Many followed Kripke and Putnam in accepting that terms like 'water' or 'tiger' refer not by description, but rather to 'whatever shares the same nature' as members of some ostended sample. But there has been far more debate about whether (as Putnam (1975, 160–165) alleges) the same theory of reference applies to artifactual terms such as 'pencil' or to social kind terms such as 'pediatrician'.

Olivero plausibly argues that many such debates can be better seen not as debates about what the 'real meaning' of the term is, or what the 'real nature' of pencils or pediatricians is, but rather as what I (2016), following David Plunkett and Tim Sundell (2013) have called 'metalinguistic negotiations'. That is, in at least many of these cases, what's going on is that disputants are implicitly arguing about how the relevant term ('cat', 'pencil', 'pediatrician'…) *ought to be used.* This is closely related to a point I emphasize in *Norms and Necessity:* that the meanings for our terms are often "incomplete, open-textured and open to change and renegotiation" (2020a, 214). Many of the difficulties which arise are cases in which it has simply not been considered or decided whether, say, the term 'pencil' could still be applied in cases where these familiar things grow on trees. As Olivero emphasizes, this does not mean that such decisions can't be made rationally—instead, we can make such decisions based on a combination of pragmatic and empirical factors. Decisions may also go in different directions depending on the interests involved. For example, as I argued (2007a, 48–53; see also my 2020a, 95–105), if we had a term that we always presumed to apply to animals (say 'cat' or 'Key Sparrow') and it turns out that what we were observing was little robots, our decisions might go in different directions if our interests are in tracking the mouse-catchers around our house (in Putnam's famous case), versus if our interests are in completing a guide to ornithology (in the Key Sparrows case). But we will be able to understand such debates better, and assess them more authentically, if we present them transparently as debates about how we *ought to* go on in such unexpected cases, rather than presenting them as if we have made 'discoveries' about how such terms 'really work' or about what the metaphysical modal facts 'really are'.

To outsiders uninterested in semantics, it may seem uninteresting or unimportant to work through how a term like 'pencil' should be used, in cases in which the pencil-like objects grew on trees; or cases in which cats turned out to be 'little robots' or pediatricians were Martian spies. But Olivero nicely brings out the general importance of the idea that we can and often must make practical decisions about how our terms should be used, by connecting it to work in conceptual engineering. Questions (for example) about how our gender terms *should* be used (especially once intersex and trans people are brought into consideration), or how a term like 'parent' *should* be used, given the complex biological, social, and institutional relations that may hold between adults and children, are centrally important to how we live and organize our social and legal institutions. Such cases are clearly not a matter of a philosopher's speculation about way-out cases. Instead, they are cases that bring out the importance of thinking through how terms should be applied in such cases, while being transparent that we are not *discovering* essences or meanings, but *deciding* how we should go on, while taking relevant interests and empirical facts into consideration. As such, they bring out the way in which there is a crucial role for conceptual engineering, which can (as I argue (2020b) and (2025b)) be done in a clear and pragmatic way. Olivero's development of a 'negotiating' approach to understanding disputes about meaning is an important contribution to developing and defending this view, for a wide range of debates. And the view, in turn, is important for not only addressing philosophical issues about reference, but also to seeing the relevance of philosophy to how we think, talk, and live.

Works Cited

Carnap, R. (1950/1956). Empiricism, semantics, and ontology. Reprinted in *Meaning and necessity*, 2nd edition. University of Chicago Press (second edition 1956)..

Cumpa, J., & Bueno, O. (2021). Resisting easy inferences. *Philosophy and Phenomenological Research, 102*(3), 729–735.

Davies, D. (2004). *Art as performance*. Blackwell.

Dodd, J. (2012). Defending the discovery model in the ontology of art: A reply to Amie Thomasson on the *Qua* Problem. *British Journal of Aesthetics, 52*, 75–95.

Evans, G. (1973). The causal theory of names. *Proceedings of the Aristotelian Society Supplementary Volume, 47*, 187–225.

Halliday, M. (2009). *The essential Halliday.* Ed. Jonathan Webster. Continuum.

Ingarden, R. (1989). *The ontology of the work of art.* Translated by Raymond Meier. Ohio University Press.

Lewis, D. K. (1986). *On the plurality of worlds*. Blackwell.

Parsons, T. (1980). *Non-existent objects*. Yale University Press.

Plunkett, D., & Sundell, T. (2013). Disagreement and the semantics of normative and evaluative terms. *Philosophers Imprint, 13*(23), 1–37.

Price, H. (2009). Metaphysics after Carnap: The ghost who walks? In D. Chalmers, D. Manley, & R. Wasserman (Eds.), *Metametaphysics*. Oxford University Press.

Putnam, H. (1975). The meaning of 'meaning'. In *Mind, language and reality: Philosophical papers* (Vol. 2, pp. 215–271). Cambridge University Press.

Restall, G. (2005). Multiple conclusions. In P. Hájek, L. Valdés-Villanueva, & D. Westerståhl (Eds.), *Logic, Methodology and Philosophy of Science: Proceedings of the Twelfth International Congress* (pp. 189–205). London: KCL Publications.

Ripley, D. (2013). Paradox and failures of cut. *Australasian Journal of Philosophy, 91*(1), 139–64.

Schiffer, S. (1996). Language-created language-independent entities. *Philosophical Topics, 24*(1), 149–167.
Sider, T. (2011). *Writing the book of the world*. Oxford University Press.
Thomasson, A. L. (2003a). Speaking of fictional characters. *Dialectica, 57*(2), 207–226.
Thomasson, A. L. (2003b). Fictional characters and literary practices. *British Journal of Aesthetics, 43*(2), 138–157.
Thomasson, A. L. (2003c). Realism and human kinds. *Philosophy and Phenomenological Research, 67*, 580–609.
Thomasson, A. L. (2004). The ontology of art. In P. Kivy (Ed.), *The Blackwell guide to aesthetics* (pp. 78–92). Blackwell.
Thomasson, A. L. (2005). The ontology of art and knowledge in aesthetics. *Journal of Aesthetics and Art Criticism, 63*(3), 221–229.
Thomasson, A. L. (2006). Debates about the ontology of art: What are we doing here? In *Philosophy compass* (Vol. 1). Blackwell. www.philosophy-compass.com, http://www.blackwell-compass.com/subject/philosophy/section_home?section=phco-aesthetics
Thomasson, A. L. (2007a). *Ordinary objects*. Oxford University Press.
Thomasson, A. L. (2007b). Artifacts and human concepts. In E. Margolis & S. Laurence (Eds.), *Creations of the mind: Theories of artifacts and their representation* (pp. 52–73). Oxford University Press.
Thomasson, A. L. (2009). Fictional entities. In J. Kim, E. Sosa, & G. Rosenkrantz (Eds.), *A companion to metaphysics* (2nd ed., pp. 10–18). Blackwell.
Thomasson, A. L. (2010). Fiction, existence and indeterminacy. In J. Woods (Ed.), *Fictions and models: New essays* (pp. 109–148). Philosophia Verlag.
Thomasson, A. L. (2015). *Ontology made easy*. Oxford University Press.
Thomasson, A. L. (2016). Metaphysical disputes and metalinguistic negotiation. *Analytic Philosophy, 57*(4), 1–28.
Thomasson, A. L. (2017). What can we do, when we do metaphysics? In G. D'Oro & S. Overgaard (Eds.), *Cambridge companion to philosophical methodology*. Cambridge University Press.
Thomasson, A. L. (2017a). Metaphysics and conceptual negotiation. *Philosophical Issues, 27*, 364–382.
Thomasson, A. L. (2017b). Husserl on essences: A reconstruction and rehabilitation. *Grazer Philosophische Studien, 94*, 436–59.
Thomasson, A. L. (2020a). *Norms and necessity*. Oxford University Press.
Thomasson, A. L. (2020b). A pragmatic method for normative conceptual work. In H. Capellen, D. Plunkett, & A. Burgess (Eds.), *Conceptual engineering and conceptual ethics* (pp. 435–458). Oxford University Press.
Thomasson, A. L. (2020c). What do easy inferences get us? *Philosophy and Phenomenological Research*. https://doi.org/10.1111/phpr.12769
Thomasson, A. L. (2021a). Philosophy as conceptual engineering. *The Philosopher, 109*(3), 7–14.
Thomasson, A. L. (2021b). Conceptual engineering: When do we need it? How can we do it? *Inquiry, 2021*, 1–26. https://doi.org/10.1080/0020174X.2021.2000118
Thomasson, A. L. (2022). How should we think about linguistic function? *Inquiry*. https://doi.org/10.1080/0020174X.2022.2074886
Thomasson, A. L. (2023a). How it all hangs together. In M. Garcia (Ed.), *Thomasson on ontology* (Philosophers in depth) (pp. 9–38). Palgrave.
Thomasson, A. L. (2023b). A neo-pragmatist approach to modality. In J. Gert (Ed.), *Neo-pragmatism: Interventions in first-order philosophy*. Oxford University Press.
Thomasson, A. L. (2025a). The ontology of art: Keeping it real, keeping it easy. In A. King (Ed.), *Art and philosophy*. Oxford University Press.
Thomasson, A. L. (2025b). *Rethinking metaphysics*. Oxford University Press.
Walton, K. (1990). *Mimesis as make-believe*. Harvard University Press.
Yablo, S. (1999, November). The strange thing about the figure in the bathhouse. Review of Amie L. Thomasson, Fiction & Metaphysics. *Times Literary Supplement, 5*.
Zalta, E. (1983). *Abstract objects*. Reidel.

Index

M. Sambrotta (ed.), *Metaphysics Today: In Conversation with Amie Thomasson*, Synthese Library 512, https://doi.org/10.1007/978-3-031-98468-6

MIX
Papier aus verantwortungsvollen Quellen
Paper from responsible sources
FSC® C105338

If you have any concerns about our products,
you can contact us on
ProductSafety@springernature.com

In case Publisher is established outside the EU,
the EU authorized representative is:
Springer Nature Customer Service Center GmbH
Europaplatz 3, 69115 Heidelberg, Germany

Printed by Libri Plureos GmbH
in Hamburg, Germany